Explainable AI for Cybersecurity

Zhixin Pan • Prabhat Mishra

Explainable AI for Cybersecurity

 Springer

Zhixin Pan
Department of Computer and Information
Science and Engineering
University of Florida
Gainesville, FL, USA

Prabhat Mishra
Department of Computer and Information
Science and Engineering
University of Florida
Gainesville, FL, USA

ISBN 978-3-031-46481-2   ISBN 978-3-031-46479-9   (eBook)
https://doi.org/10.1007/978-3-031-46479-9

This Springer imprint is published by the registered company Springer Nature Switzerland AG
The registered company address is: Gewerbestrasse 11, 6330 Cham, Switzerland

Paper in this product is recyclable.

Preface

We are living in the age of smart computing systems powered by artificial intelligence as well as hardware (electronics) and software (applications). We can visualize the smart computing when we are using smartphones or laptops. Even when we do not feel its presence, smart computing is in action. For example, cars and airplanes have tons of electrics and millions of lines of software code to make our journey comfortable. Security is a major concern in designing these systems. Recent attacks have demonstrated that an attacker can exploit hardware and software vulnerabilities in a wide variety of systems. In order to design trustworthy systems, it is crucial to identify and mitigate both hardware and software vulnerabilities. There are recent efforts in utilizing artificial intelligence for defending against cybersecurity attacks.

Artificial Intelligence (AI) is successful in achieving human-level performance in various fields using machine learning (ML) algorithms. However, ML models lacks the ability to explain an outcome due to its black-box nature. Explainable AI has received significant attention due to its ability to provide insights into the decision-making process. This enables efficient detection and localization of malicious software attacks (e.g., malware and ransomware), hardware attacks (e.g., hardware Trojans), as well as combinational attacks (e.g., Spectre and Meltdown attacks).

This book provides a comprehensive overview of security vulnerabilities and state-of-the-art countermeasures using explainable AI. Specifically, it describes how explainable AI can be effectively used to design secure and trustworthy systems. It also explores hardware acceleration of explainable AI using field programmable gate array (FPGA), graphics processing units (GPU), and tensor processing units (TPU). The presentation of topics has been divided into six categories with each category focusing on a specific aspect of the big picture. A brief outline of the book is provided below.

1. **Introduction to Cybersecurity and Explainable AI:** The first part of the book includes two introductory chapters on cybersecurity and explainable AI.

- Chapter "Cybersecurity Landscape for Computer Systems" provides an overview of security vulnerabilities as well as defense mechanisms against cybersecurity attacks.
- Chapter "Explainable Artificial Intelligence" introduces various machine learning algorithms and describes popular explainable AI methods.

2. **Detection of Software Vulnerabilities:** The second part of the book focuses on detection of malicious software attacks.

- Chapter "Malware Detection Using Explainable AI" presents a malware detection and localization framework using explainable AI.
- Chapter "Spectre and Meltdown Detection Using Explainable AI" describes an efficient framework for detection and localization of Spectre and Meltdown Attacks using explainable AI.

3. **Detection of Hardware Vulnerabilities:** The third part of the book deals with detection of hardware Trojans.

- Chapter "Hardware Trojan Detection Using Reinforcement Learning" describes hardware Trojan detection using reinforcement learning.
- Chapter "Hardware Trojan Detection Using Side-Channel Analysis" provides an overview of hardware Trojan detection techniques using side-channel analysis.
- Chapter "Hardware Trojan Detection Using Shapley Ensemble Boosting" utilizes Shapley ensemble boosting for detection of hardware Trojans.

4. **Mitigation of AI Vulnerabilities:** The fourth part of the book looks at mitigation of AI vulnerabilities.

- Chapter "Mitigation of Adversarial Machine Learning" explores mitigation techniques to design robust ML models against adversarial attacks.
- Chapter "AI Trojan Attacks and Countermeasures" presents various AI Trojan attacks and effective countermeasures.

5. **Acceleration of Explainable AI:** The fourth part of the book explores hardware-based acceleration of explainable AI.

- Chapter "Hardware Acceleration of Explainable AI" describes hardware acceleration of explainable AI using FPGA and GPU.
- Chapter "Explainable AI Acceleration Using Tensor Processing Units" presents TPU-based acceleration of explainable AI.

6. **Conclusions and Future Directions:** The last part concludes the book with a summary and discussion on future directions.

- Chapter "The Future of AI-Enabled Cybersecurity" concludes the book with an executive summary as well as discussion on designing future-proof trustworthy systems.

We hope you enjoy reading this book and find the information useful for designing secure and trustworthy systems.

Gainesville, FL, USA   Zhixin Pan
August 31, 2023   Prabhat Mishra

Acknowledgements

This book is the result of a decade long academic research and industrial collaborations. The book includes the hardware security verification techniques and insights that resulted from the Ph.D. dissertation of Dr. Zhixin Pan. We would like to acknowledge our sponsors for providing the financial support to enable this research work. This work was partially supported by National Science Foundation (CCF-1908131). We would like to acknowledge that Emma Andrews (University of Florida) has contributed a book chapter on explainable artificial intelligence (AI). We also acknowledge that Sahan Sanjaya (University of Florida) has contributed a book chapter on hardware acceleration of explainable AI.

Contents

Part III   Detection of Hardware Vulnerabilities

Hardware Trojan Detection Using Reinforcement Learning

Hardware Trojan Detection Using Side-Channel Analysis

Hardware Trojan Detection Using Shapley Ensemble Boosting

Acronyms

ABS	Artificial Brain Stimulation
ADC	Analog-to-Digital Converter
AI	Artificial Intelligence
ASIC	Application-Specific Integrated Circuit
ASR	Attack Success Rate
ATPG	Automatic Test Pattern Generator
AVS	Anti-Virus software
BFD	Best-Fit-Decrease
BNN	Bayesian Neural network
BP	Back Propagation
CLB	Configurable Logic Block
CMM	Cascaded Matrix Multiplication
CNN	Convolutional Neural Networks
DAC	Digital-to-Analog Converter
DoS	Denial of Service
DDoS	Distributed Denial-of-Service
DFT	Discrete Fourier Transform
DNN	Deep Neural Networks
DRAM	Dynamic Random Access Memory
DSP	Digital Signal Processor
DT	Decision Tree
EM	Electromagnetic emanation
ETB	Embedded Trace Buffer
FC	Fully-Connected Layer
FFD	First-Fit-Decrease
FNR	False Negative Rate
FOA	Focus of Attention
FP	Forward Pass
FPGA	Field Programmable Gate Arrays
FPR	False Positive Rate
FSM	Finite State Machine

GPU	Graphics processing units
HBM	High Bandwidth Memory
HLS	High-Level Synthesis
HPC	Hardware Performance Counters
HT	Hardware Trojan
IC	Integrated Circuit
I/O	Input/Output
IOR	Inhibition of Return
IP	Intellectual Property
LIME	Local Interpretable Model-Agnostic Explanations
LR	Linear Regression
LRP	Layer-wise Relevance Propagation
LSTM	Long Short-Term Memory
LUT	Look Up Table
MAC	Multiplication-and-Addition
MLaaS	Machine Learning as a Service
MLP	Multi-Layer Perceptron
MXU	Matrix Multiply Unit
NLP	Natural Language Processing
PCI	Peripheral Component Interconnect
QoS	Quality-of-Service
RE	Reverse Engineering
RF	Random Forest
RNN	Recurrent Neural Network
ROs	Ring Oscillators
RL	Reinforcement Learning
RTL	Register Transfer Level
SCA	Side-Channel Analysis
SCOAP	Sandia Controllability/Observability Analysis Program
SCV	Supply Chain Vulnerability
SEB	Shapley Ensemble Boosting
SGD	Stochastic Gradient Descent
SHAP	Shapley Values
SIMT	Single Instruction Multiple Thread
SM	Streaming Multiprocessors
SoC	System-on-Chip
SPEC	Standard Performance Evaluation Corporation
SVM	Support Vector Machine
TPU	Tensor Processing Units
UART	Universal Asynchronous Receiver/Transmitter
VM	Virtual Machine
WTA	Winner-Take-All
XAI	Explainable Artificial Intelligence
ZSL	Zero-Shot Learning

Part I
Introduction

Cybersecurity Landscape for Computer Systems

1 Introduction

A vast majority of our daily activities involve interactions with the electronic systems, such as desktops, laptops, and smartphones. Security and privacy of such interactions rely on the trustworthiness of the underlying computer systems, which consists of hardware as well as software. In order to design trustworthy computer systems, we need to first identify the sources of hardware and software vulnerabilities. Next, we need to implement effective detection and mitigation methods to defend against cybersecurity attacks.

Figure 1 shows that the hardware is the brain behind the computing systems. A wide variety of hardware components are assembled as *integrated circuits* (ICs). For example, a System-on-Chip (SoC) consists of diverse Intellectual Property (IP) cores including processor, memory, network-on-chip, controllers, converters, and input/output devices. Unlike microcontroller based designs in the past, even resource constrained Internet-of-Things (IoT) devices nowadays incorporate one or more complex SoCs. Drastic increase in SoC complexity has led to significant increase in hardware design and validation complexity. Semiconductor companies utilize global supply chain during SoC design and manufacturing to reduce cost and meet time-to-market constraints. Figure 2 shows a typical SoC design flow involving multiple third-party IP vendors. Unfortunately, reliance on third-party IPs raises hardware security concerns [14, 20, 58, 59]. For example, a hardware IP gathered from a potentially untrusted vendor may come with malicious implants (e.g., hardware Trojans), backdoor for information leakage, or other integrity issues. Based on Common Vulnerability Exposure (CVE-MITRE) estimates, if hardware-level vulnerabilities are removed, the overall system vulnerability will reduce by 43% [1]. It is critical to detect and remove these vulnerabilities for designing secure and trustworthy systems.

© The Author(s), under exclusive license to Springer Nature Switzerland AG 2023
Z. Pan, P. Mishra, *Explainable AI for Cybersecurity*,
https://doi.org/10.1007/978-3-031-46479-9_1

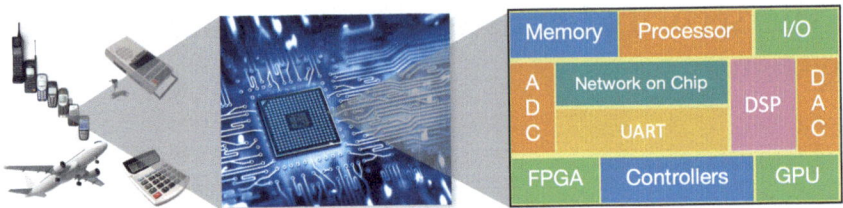

Fig. 1 An example System-on-Chip (SoC) consisting of diverse components, including processor, digital signal processor (DSP), memory, controllers, analog-to-digital converter (ADC), digital-to-analog converter (DAC), graphics processing units (GPU), filed programmable gate arrays (FPGA), universal asynchronous receiver/transmitter (UART), input/output (I/O), etc.

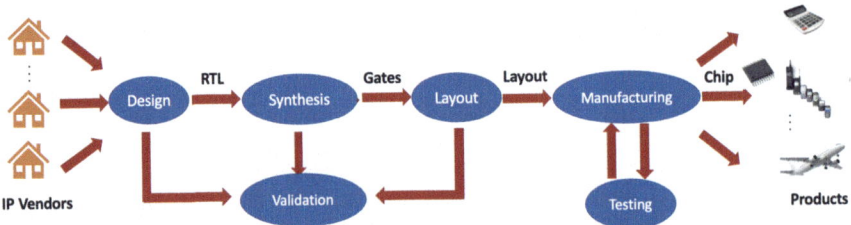

Fig. 2 System-on-Chip (SoC) design flow involving third-party Intellectual Property (IP) cores. Various IPs are integrated to produce register-transfer level (RTL) design. The synthesis step transforms the RTL design into a gate-level netlist. The next step produces the layout, which is used during manufacturing to produce integrated circuits. Pre-silicon as well as post-silicon validation are essential to ensure functional correctness as well as non-functional requirements

Computer systems also utilize software components, a set of programs written in specific programming languages. These software components play an important role to interact between users and the hardware components. Malicious software, popularly known as malware, is widely acknowledged as a serious threat to modern computing systems. There are a wide variety of malware, including adware, spyware, ransomware, botnets, etc. Malware can harm a computer, server, or computer network and cause damage to the target system. The portability of malware also enables them to proliferate across various platforms at an alarming rate. Our reliance on diverse smart devices as well as open-source applications provided by unverified third-party developers have created a perfect storm for privacy leakage through malware-infected embedded systems. A recent study reported that over 33,412,568 unique malicious files were detected in 2020 with an average revenue loss of $2.6 million per organization [48]. Clearly, there is an urgent need to develop efficient malware detection techniques to enable trustworthy computing.

Due to the increasing importance of security, there are a wide variety of solutions for defending against hardware and software attacks. Although these approaches provide promising results, they commonly suffer from expensive resource cost, vulnerability against obfuscation techniques, and reliance on human expert knowledge. While recent solutions utilize Machine Learning (ML) techniques for vulnerability detection across various application domains, they have several practical limitations [66]. Due to the inherent inefficiency of ML algorithms, they are usually not applicable in real-time systems. Moreover, since ML algorithms are commonly implemented as software programs, they are also vulnerable toward adversarial attacks. In order to design trustworthy systems, it is necessary to defend against exploitation of various vulnerabilities, including hardware, software, and machine learning models. In this book, we investigate the effectiveness of explainable AI to defend against such attacks.

The remainder of this chapter is organized as follows. Section 2 provides an overview of security vulnerabilities in hardware (Sect. 2.1), software (Sect. 2.2) as well as machine learning models (Sect. 2.3). Section 3 describes efficient detection and mitigation methods using machine learning algorithms. Finally, Sect. 4 summarizes this chapter.

2 Cybersecurity Vulnerabilities

In this section, we describe a wide variety of security vulnerabilities in computer systems that an attacker can exploit. We first provide an overview of hardware vulnerabilities. Next, we describe software vulnerabilities. Finally, we outline adversarial attacks on machine learning models.

2.1 Hardware Vulnerabilities

In this section, we provide a brief overview of various hardware vulnerabilities, including malicious implants (hardware Trojans), supply chain vulnerability, reverse engineering, and side-channel leakage.

2.1.1 Malicious Implants (Hardware Trojans)

Hardware Trojan (HT) [84] is a malicious hardware modification that can leak secret information, degrade the performance of the system, or cause denial-of-service. It is a malicious modification of the target integrated circuit (IC) with two critical parts: trigger and payload. When the trigger is activated, the payload enables the malicious activity [24]. For example in Fig. 3, when the output of the trigger logic is true, the output of the payload XOR gate will invert the expected output. The trigger is typically created using a combination of rare events (such as rare signals or rare transitions) to stay hidden during normal execution. The payload represents the malicious impact that an HT will inflict to the target design, commonly resulting in information leakage or erroneous execution. HT can be introduced at various components, including computation modules (e.g., IPs) and

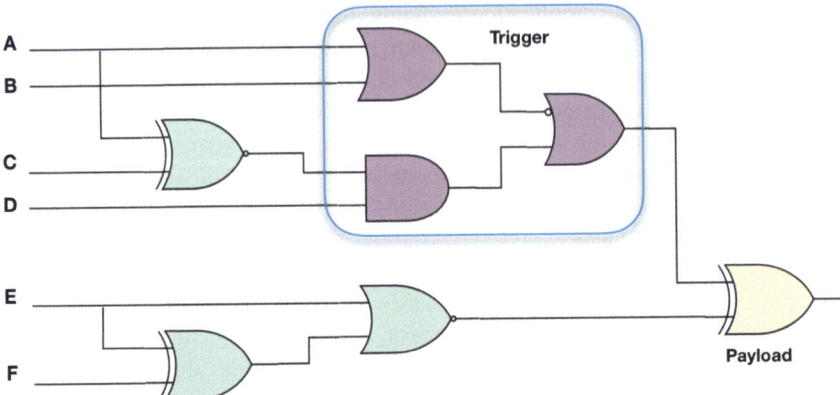

Fig. 3 An example hardware Trojan constructed by a trigger logic (purple gates). Once the trigger condition is satisfied, the payload (yellow XOR gate) will invert the expected output. The gates of the original design are shown in green color

communication channels (e.g., bus) [11, 21, 22, 26, 62, 81]. A major challenge for Trojan identification is that Trojans are usually stealthy, as they are designed in a way that they can be activated under very rare conditions [10]. Due to this stealthy nature, it is infeasible to detect them using traditional functional validation methods [80].

2.1.2 Supply Chain Vulnerability

Supply chain vulnerability (SCV) can be broadly categorized into three fundamental threats: counterfeiting, overbuilding, and recycling. IC counterfeiting is a serious problem that arises from the global semiconductor supply chain [39]. Some manufacturers use recycled ICs instead of genuine ones for reasons like out of stock or lower price. Counterfeit ICs with poor quality are usually overused, which degrade the quality of the manufactured products and lead to early product failures [32]. Counterfeiting of ICs has become a major challenge due to the difficulty of detection and the lack of effective avoidance mechanisms [38]. Most of the existing approaches require costly rework and significant time overhead. IC overbuilding is a situation where the foundry manufactures more ICs than required by the IC designer. During the flow of supply chain, an attacker with access to these ICs can steal and claim ownership of the extra ICs and sell them illegally. Specifically, IC overbuilding attacks are launched by illegally copying or stealing authentic blueprints of SoC during the design, synthesis, or production phases and result in illegal sales in the market, as demonstrated in [8].

2.1.3 Reverse Engineering

Reverse engineering (RE) of SoC [25] is an information invasion technique achieved by exploiting the backdoors in an SoC either by faulty design or malicious implants [57]. A successful reverse engineering attack enables the adversary to uncover the IC design and extract its gate-level netlist to further infer its functionality [78]. This will lead to the revealing of inner details about the design, where the attacker can steal the intellectual property or even improve it on their own product to achieve illegal advantage. The threat of reverse engineering is discussed in detail in [4].

2.1.4 Side-Channel Leakage

Side-channel vulnerabilities arise from the fact that electronic devices inevitably produce physical emanation during execution, including but not limited to execution time, power consumption, path delay, and electromagnetic emanation. These physical signatures can unintentionally reveal secret information from the device. For example, timing attack is a typical approach to abuse the side-channel vulnerability, since timing information can be exploited to reveal memory information. Assuming that there is a private array in the memory with one pre-recorded entry in the cache, while the attacker wants to know the actual index of it. This can be achieved by traversing the entire array and measuring the access time. The location with the shortest access time is the target one due to the huge difference of access time between cache and memory, as shown in Fig. 4. This cache-based side-channel attack is widely applied in the famous Spectre and Meltdown attacks [63].

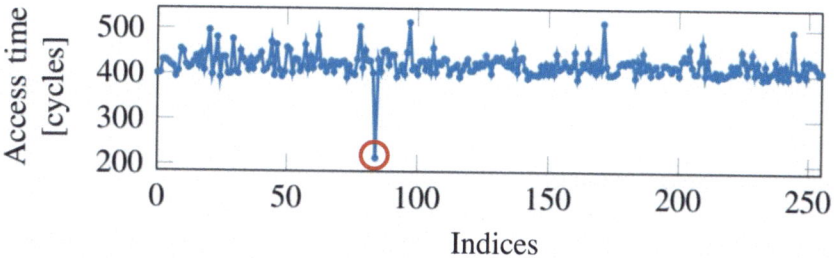

Fig. 4 An example of cache-based side-channel attack. By traversing the entire array and recording the access time, the adversary can lock on the data corresponding to the index with the shortest access time, since it is likely to be pre-stored in the cache

2.2   Software Vulnerabilities

Malicious software, popularly known as malware, is widely acknowledged as a serious threat to modern computing systems. In this section, we provide a brief overview of various malware attacks, including malware attacks, ransomware attacks, and Spectre and Meltdown attacks.

2.2.1   Malware Attacks

Malware, short for "malicious software," refers to any type of software specifically designed to harm, exploit, or compromise computer systems, networks, or users' data. Malware is typically created with malicious intent, such as stealing sensitive information, disrupting computer operations, or gaining unauthorized access to systems. There are various types of malware, including but not limited to viruses, worms, spyware, and botnets. Malware can spread through various vectors, including email attachments, malicious websites, infected software downloads, and compromised devices. Malware attacks can lead to financial losses due to theft of sensitive information (such as credit card details and bank account credentials), ransom payments, and costs associated with mitigating the attack and recovering from its effects. At the same time, malware can also be used to infiltrate and exfiltrate sensitive data, leading to data breaches. This can result in the exposure of personal and confidential information, causing harm to individuals and damaging the reputation of organizations.

A key problem to note is that malware attacks are constantly evolving, with attackers adapting their tactics to bypass security measures. As a result, cybersecurity professionals and organizations need to stay vigilant, update their security practices, use advanced threat detection tools, and continuously educate their users about safe online behavior. Collaboration between security experts, law enforcement agencies, and technology companies is crucial in the ongoing battle against malware.

2.2.2   Ransomware Attacks

Ransomware is one of the most widespread malware cyber-attack that holds the victim's data hostage by surreptitiously encrypting the data on a user's computer to make it unavailable and only decrypts the data after the user pays a ransom. An example of the infamous WannaCry Ransomware is shown in Fig. 5 [83].

Ransomware generally comes in two primary forms: Crypto ransomware and Locker ransomware. The crypto ransomware attacks the victim's machine by discreetly searching for documents in the victim's machine and then encrypting them. Victims can get back access to their documents only after paying a ransom and receiving the decryption keys from the attackers. Usually, crypto ransomware do not

Fig. 5 An example of ransomware infection. Image credit [83]

encrypt the entire physical drive but targets user-generated files that have specific file extensions, such as pdf, jpg, and doc, which typically contain valuable and personal user data. On the other hand, the locker ransomware locks the user's computer, thus preventing the user from accessing their files. A distinguishing feature of the locker ransomware strain is that once the user's computer and files have been taken hostage, the ransomware announces its presence to the user by displaying a ransom note on the user's computer screen demanding a ransom payment in exchange for the return of the user's files. Only after a successful payment, the user is granted access back to the affected machine. CryptoWall is an example of locker ransomware that has been described as the most destructive ransomware threat on the Internet as it is programmed to run on both 32-bit and 64-bit machines, thus increasing its chances of infection.

2.2.3 Spectre and Meltdown Attacks

Modern processors support advanced architectural features for performance improvement, such as branch prediction, out-of-order, and speculative execution, as shown in Fig. 6. Spectre and Meltdown attacks exploit these architectural features to attack security guarantees provided by an operating system. The operating

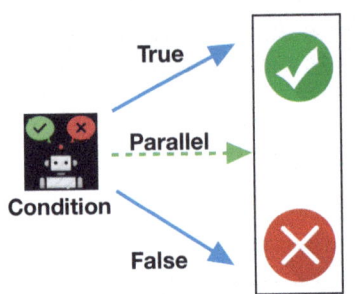

The chip computes both statements in parallel before condition checking.

Out-of-order execution to avoid waiting caused by fetching the next instruction.

Fig. 6 Typical strategies applied in speculative execution

system has one of the most fundamental security requirements—it must prevent user programs from accessing the memory locations of the kernel or any other programs. Once a user program tries to perform illegal access, the CPU will detect the permission violation during the execution and throw an exception leading to the termination of current program. However, during this permission checking and scene clearing process, the information about accessing target is retained in the cache. These are inherent vulnerabilities in most of modern chips, which can be exploited by attackers to reveal kernel memory information. Specifically, a simple template of Meltdown attack code is shown in Listing 1.

```
mov rax byte[x]     // illegal access
shl rax 0xC         // page alignment
mov rcx rbx[rax]    // probe data
```

Listing 1 An example Meltdown attack

In this example, "rax," "rbx," and "rcx" are register names. Here, "byte[x]" is a private memory location; illegal access to this location shall raise exception during execution. The left shift by 12 bits in the second instruction enables multiplication of "rax" by the page size (4096). Ideally, "rax" should be cleared before executing the subsequent instructions. However, due to the speculative execution property, the second and third instructions will be partially executed before the exception handling takes effect. Also, according to the modern cache designs, if "rax" is not in the cache, the CPU will bring it into the cache to hide the latency of subsequent accesses. Although "rax" will be cleared by exception handling, the cache will not be flushed immediately. Therefore, the information of the latest illegal access is temporarily stored in the cache. An attacker can restore this address by a *cache-based side-channel attack*. This is achieved by simply traversing the entire array headed by "rbx" and measuring the access time – the page with the shortest access time is the one addressed by "rax," thereby this kernel value is obtained.

Spectre attack is very similar but it exploits branch prediction instead, which is shown in Listing 2. Obviously, if index "x" is out of range, the second line should not be executed. Due to branch prediction scheme, it will still be "pre-executed." Once this pre-execution occurs, it will inevitably leave traces in the cache, where the same cache-based side-channel attack can be used. Compared to Meltdown, Spectre is more dangerous since it has a wider attack range [44].

```
if(x < arr1_size );    // boundary check
y = arr2[arr1[x]*4096];  // array access
```

Listing 2 An example Spectre attack

2.3 Malicious Attacks on Machine Learning Models

Machine Learning (ML) algorithms are widely used in cybersecurity vulnerability detection due to its outstanding performance in both supervised and unsupervised scenarios. The flexibility of ML models also enables their different variations to be successfully employed in diverse applications. Since most of ML algorithms are implemented in the software layer, they are also vulnerable toward malicious attacks. In this section, we describe two popular attacks on machine learning models.

2.3.1 Adversarial Attacks

The adversarial attack observed by Szegedy et al. [77] revealed the vulnerability of most existing deep neural networks (DNNs) against adversarial examples. The difference between adversarial samples and the original sample can hardly be distinguished by naked eyes but will lead the model to make an incorrect prediction with high confidence. As shown in Fig. 7, a human-invisible noise is added to the input traffic sign image. While a pre-trained network can successfully recognize the original input as a stop sign, the same network will incorrectly classify it as a yield sign if the input is perturbed with well-crafted noise [87]. Existing mitigation techniques against adversarial attacks like model retraining [6] and pruning [17] often cause significant performance degradation. In order to design robust DNNs, it is necessary to develop efficient defense against adversarial attacks.

2.3.2 AI Trojan Attacks

Aside from adversarial attacks, ML algorithms are also vulnerable toward malicious implants. ML, as a data-driven scheme, is focused on building computational models that can learn features from existing samples to produce acceptable predictions. However, ML models are computationally expensive to train, requiring huge amount of computation resources. To reduce cost, many complex applications rely on

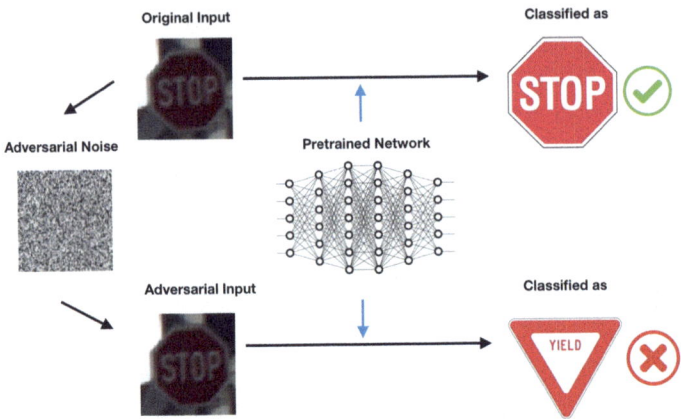

Fig. 7 An example adversarial attack on autonomous driving: a stop sign miss-classified as a yield sign due to noise

outsourcing the training procedure to the cloud service or rely on pre-trained models. This process is referred as *Machine Learning as a Service (MLaaS)*. While MLaaS provides specific advantages, it also provides adversaries with opportunities to launch backdoor attacks toward ML models, popularly known as AI Trojans in computer vision domain. AI Trojans and hardware Trojans are similar from several perspectives. (1) Both of them are malicious implants consisting of a rare trigger and a small payload. The trigger represents a rare condition, and therefore, it is hard to activate using traditional validation methodology. The payload is typically very small compared to the design (model), and therefore, it is hard to detect. (2) The functionality of the infected circuit (or backdoored ML model) is not affected until the adversary applies certain inputs to activate the Trojan trigger. (3) They can be inserted by a rogue employee or an adversary involved in any of the third-party service (e.g., MLaaS for AI Trojans or IP design/synthesis/fabrication for hardware Trojans). In spite of the above similarities, they have one major difference. While the primary objective of AI Trojans is to mispredict (incorrect execution), hardware Trojans can lead to information leakage, incorrect execution, denial-of-service, or other unintended consequences. Although Bayesian neural network is promising to provide robustness against AI Trojan attacks, it fails to mitigate sophisticated AI Trojans [65].

3 Detection of Security Vulnerabilities

In this section, we survey the recent approaches for detection and mitigation of hardware as well as software vulnerabilities using machine learning (ML) algorithms.

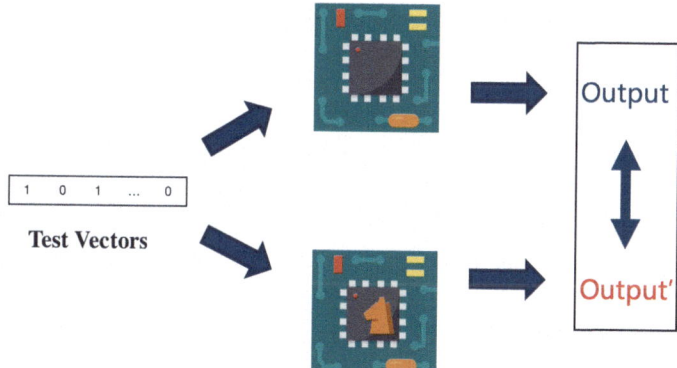

Fig. 8 An overview of simulation-based validation of security vulnerabilities

3.1 Detection of Malicious Hardware Attacks

There are a wide variety of detection techniques against malicious attacks on computer hardware. In this section, we focus on some of the ML-based detection methods.

3.1.1 Simulation-Based Validation Using Machine Learning

Simulation-based validation approaches focus on test generation to activate hardware vulnerabilities. As shown in Fig. 8, the presence of a malicious implant (hardware Trojan) can be detected if we can simulate with a suitable test vector. If the test vector activates the trigger of the Trojan, comparison of the simulation output with the expected output will reveal the presence of a Trojan in the implementation. Simulation-based vulnerability analysis starts with producing test vectors, which will be fed to the target design. By observing the outputs and comparing with the golden design, it can reveal important clues for possible vulnerabilities.

Random test generation was widely explored for simulation-based approaches due to its simplicity. However, there is no guarantee for activating stealthy Trojans using millions of random or constrained-random tests. MERO [13] proposed a statistical test generation scheme, which adopts the N-detect idea [70] to achieve better coverage. The key insight is that if all rare signals are activated for at least N times, it is likely to activate the rare trigger conditions when N is sufficiently large. The left side of Fig. 9 shows an overview of MERO. It starts with random test generation followed by a brute-force process of flipping bits to increase the number of rare values being satisfied. It provides promising result for small benchmarks, but it introduces long execution time and scalability concerns, making it unsuitable for large benchmarks [55].

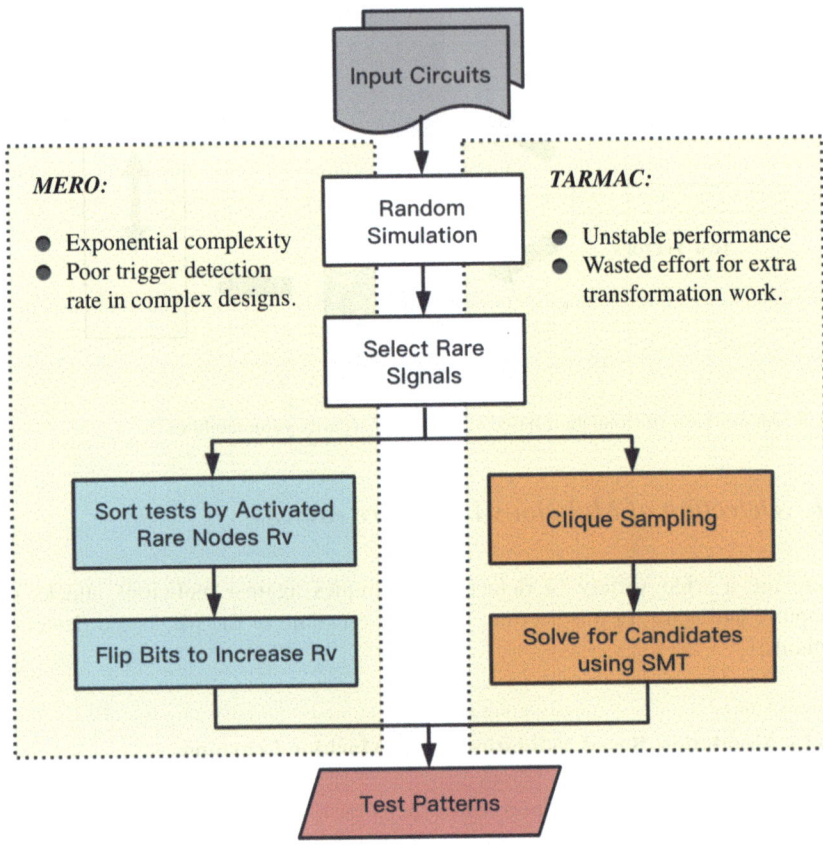

Fig. 9 Overview of state-of-the-art test generation techniques for detecting hardware Trojans: MERO [13] and TARMAC [55]

To address these issues, Lyu et al. proposed TARMAC [55] as shown on the right side of Fig. 9. Like MERO, TARMAC also starts with random simulation to identify rare signals in the netlist. Next, it maps the design to a satisfiability graph and converts the problem of satisfiability into a clique cover problem, where the authors use an SMT solver [61] to generate test patterns for each maximal clique. Although TARMAC performs significantly better than MERO in evaluated benchmarks, its performance is unstable. This is due to the fact that TARMAC relies on random clique sampling, making its performance dependent on the quality of sampled cliques. A similar runtime approach has been proposed for many-core platforms [45]. These approaches can achieve high accuracy (80%) in detecting Trojans. However, they also have a high false positive rate as they might mark many benign components as suspicious. The accuracy of such approaches has been improved by Pan et al. [64] utilizing improved feature selection and use of reinforcement learning models to reduce false positive rate. The efficiency of this framework is improved in [88] by reducing the model size using ensemble boosting. In [89], the authors further reduce the detection overhead by removing the golden-circuit requirement with the help of zero-shot learning.

3.1.2   Side-Channel Analysis Using Machine Learning

The field of side-channel analysis (SCA) has gained significant attention for hardware vulnerability analysis [50]. SCA consists of two major steps: execution using suitable input (test) patterns and comparison with the expected (golden) side-channel signatures. Security experts create test vectors to trigger side-channel signatures, which can be used to evaluate the quality of the SoC or detect potential vulnerabilities. Although both SCA and simulation-based validation (SV) rely on test generation and comparison with golden design, there are few fundamental differences.

- Unlike SV, which relies on functional values at output ports, SCA relies on side-channel signatures such power consumption, path delay, electromagnetic emanation, etc.
- Detection of tiny Trojans is difficult using SCA, since the side-channel footprint of the tiny Trojan can easily hide in environmental noise or manufacturing process margins. This is not a problem for SV—if the Trojan can be activated, there will be mismatch in the expected outputs (assuming the Trojan alters the functionality).
- It is hard to activate Trojan trigger during SV, since it is infeasible to simulate all possible input (test) patterns in a large design (exponential number of test vectors). This is not a problem in SCA, since it does not rely on the activation of the Trojan trigger.

SCA is widely adopted in scenarios where the model is poorly known [53], since the task can be approximated to a profiling phase. SCA is typically used for hardware Trojan detection in integrated circuits by analyzing various side-channel signatures, such as timing, power, and path delay. An overview of side-channel attacks and countermeasures is presented in Fig. 10. We briefly describe side-channel analysis using dynamic power (current) [34, 54, 56], path delay [67], and electromagnetic emanation [28].

Side-Channel Analysis Using Dynamic Current Dynamic current, specifically related to dynamic power consumption, can be a valuable source of information in hardware Trojan detection. Each integrated circuit has a unique power consumption profile based on its architecture, logic, and operations. By monitoring the dynamic current consumption of a device during its operation, the baseline power consumption profile can be collected for recognizing legitimate behavior. Hardware Trojans often introduce subtle changes in the device's behavior, which can affect the total power consumption. By continuously monitoring the dynamic current and comparing it to the baseline profile, it is possible to detect the anomalies or

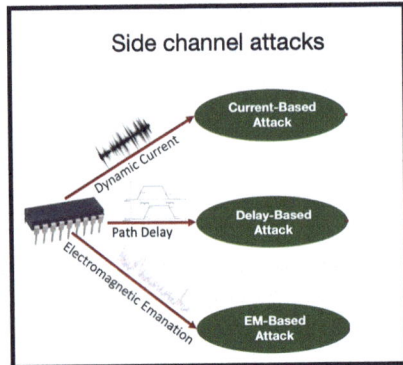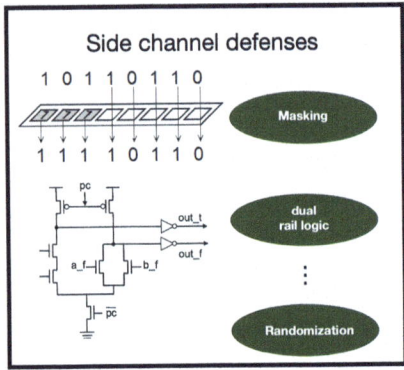

Fig. 10 A brief overview of commonly applied side-channel attacks and defenses

deviations that could indicate the presence of a hardware Trojan. A typical workflow of hardware Trojan detection using dynamic current involves the following steps:

- **Data Collection**: Users gather the integrated circuit (IC) or device to be tested, set up the measurement equipment to monitor the dynamic current consumption of the IC during its normal operation, and execute various workloads or operations on the IC to cover a wide range of scenarios and behaviors.
- **Baseline Power Consumption Profiling**: Collect dynamic current data while the IC is performing legitimate operations. The collected data are aggregated and analyzed to create a baseline power consumption profile for normal behavior.
- **Anomaly Detection**: The real-time dynamic current consumption is recorded and compared to the baseline profile to identify deviations or anomalies in the power consumption patterns that do not align with the baseline. Various techniques, such as statistical inference, behavior analysis, and machine learning, can be applied here.
- **Threshold Setting**: To improve accuracy and reduce false positive rate, appropriate thresholds are required to be established based on the statistical properties of the baseline data. The thresholds are then fine-tuned to reduce false positives while maintaining high sensitivity to Trojan detection.

Dynamic current-based analysis can be a part of the overall hardware Trojan testing and validation process. By subjecting the device to different workloads and monitoring its power consumption, a real-time monitoring framework can be implemented to identify irregularities that might indicate the presence of Trojans.

Side-Channel Analysis Using Path Delay Path delay-based side-channel analysis leverages the fact that malicious implants can affect the timing delays of specific paths within the chip. These timing delays refer to the time it takes for a signal to travel from one point to another within the chip. Hardware Trojans introduce intentional delays or variations in these paths, which can be detected by comparing the actual delay with the expected delay based on the chip's original design. Compared to other side-channel signatures, path delay provides the following three advantages:

1. *Independence:* The delay between any gates in the design can be measured independently, which provides more comprehensive information compared to other side-channel signatures.
2. *Diversity:* Implanted Trojans can impact path delay in multiple ways. There will be an increase of propagation delay for the gates producing trigger signals, since they are connected to an extra gate, which leads to increased capacitive load. Moreover, since one XOR gate and one AND gate were inserted to deliver the payload, the path delay will always have at least two gates difference from the golden design for any paths through these inserted XOR/AND gates.
3. *Stability:* Delay-based Trojan detection techniques provide superior performance under parametric variations by leveraging statistical techniques [71]. This stability guarantees high confidence of detection results from delay-based analysis.

EM-Based Side-Channel Analysis Electromagnetic emanation (EM) has been proposed as a new approach to acquire forensically useful data from IoT devices [74]. When a computing device runs a program, any changes in a signal (wire) inevitably generates EM emissions. The observable patterns depend on the precise settings of the device. In the EM emission spectrum of the device, it serves as a clear evidence for both the hardware and software's influence, reflected by the EM emission patterns. The signals captured at runtime can be utilized as a signature for identifying abnormal behaviors caused by malicious implants.

Compared to other side-channel signatures, exploiting EM for hardware Trojan detection offers two major advantages. First, the collection of EM signals is completely non-invasive and passive, which means that the measurement process can be performed without interrupting the normal execution of the target device or causing any physical damage. Next, EM emanation from electronic devices is spontaneous, which is very difficult to prevent by hardware-based countermeasures. Therefore, hardware Trojans cannot hide using their sneaky nature if we employ EM-based side-channel analysis. However, performing successful EM-based SCA attacks on IoT devices requires domain knowledge and specialized equipment (Fig. 11).

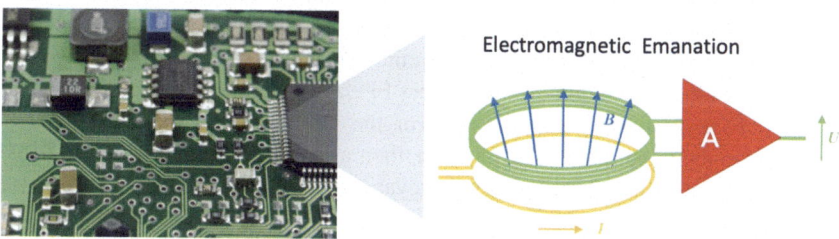

Fig. 11 The running of program produces EM emanation

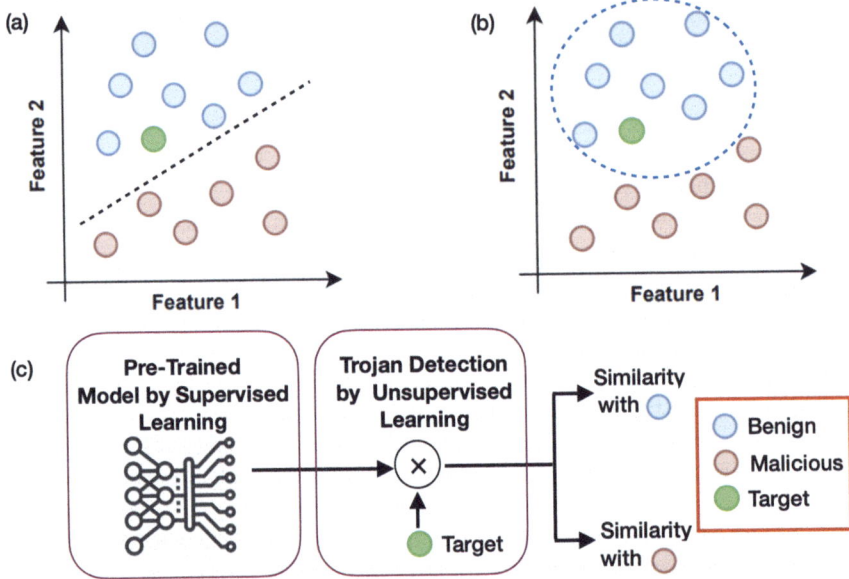

Fig. 12 The illustration of fundamental differences between three hardware Trojan detection approaches using (**a**) supervised learning, (**b**) unsupervised learning (clustering), and (**c**) zero-shot learning

3.1.3 Heuristic Analysis Using Machine Learning

There are various machine learning based methods for heuristic analysis. Broadly speaking, machine learning algorithms can be categorized into supervised learning and unsupervised learning. Figure 12a shows an example of supervised learning based classification, while Fig. 12b shows clustering based classification in unsupervised learning. There are also approaches that combine the benefits of supervised and unsupervised learning, as shown in Fig. 12c.

Heuristic Analysis Using Supervised Learning In [18], the authors target integrated circuit (IC) aging analysis by proposing a two-stage method for detecting recycled FPGAs. Both stages rely on machine learning via SVM for classification. In the first stage, they compare the frequencies of ring oscillators (ROs) distributed on the FPGAs against a golden model as the crucial feature to distinguish recycled and fresh FPGA. This is an initialization step, where they raise red flags for suspicious components in the entire circuit. While in the second phase, they performed a short aging step on the suspect FPGAs and exploited the aging speed reduction (due to prior usage) to confirm the cases marked by the first step. According to their experimental evaluation, their approach can distinguish fresh and recycled FPGAs with 100% accuracy for a variety of benchmarks.

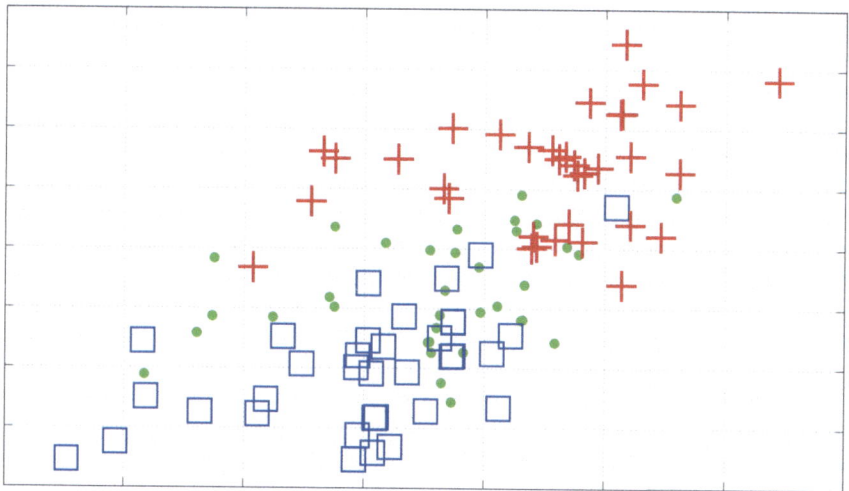

Fig. 13 IC with different aging level are mapped in 2D space, where the classification is performed using SVM to distinguish data points in 2D space. Image credit [33]

Huang et al. [33] presented an approach to detect counterfeit ICs using SVM. Instead of ROs, they first observe the distribution of process variation across devices. Next, they build a simple parametric measurement to map this circuit feature into a 2-D space. Finally, an SVM classifier performs the job of counterfeit detection, which is shown in Fig. 13. However, these approaches work well when we know the proper feature to be fed into the ML models. In other words, promising solutions for a specific scenario (e.g., countermeasure against IC counterfeiting) cannot be utilized for other types of SoC vulnerabilities.

Hasegawa [29] developed a Trojan-feature extraction algorithm at gate level and utilized it to perform hardware Trojan detection using SVM classifier. This work was also extended to similar framework with different ML models including random forest (RF) [31] and neural networks [30]. In general, they focus on hardware Trojan detection using machine learning in IC design step. They outlined 11 most effective features that reflect Trojan's impact on netlists. By using RF as the classifier, they obtained 100% true positive rate and 100% true negative rate in several Trust-HUB benchmarks.

In [12], the author estimates the statistical correlation between the signals in a design and explores how this estimation can be used in a clustering algorithm to detect the Trojan logic. Since it is based on heuristic of using statistical model to estimate the signal correlations of gate-level netlist, it neither needs the circuit to be brought to the triggering state nor the effect of the Trojan payload to be propagated and observed at the output. This idea was inherited in [51], where the authors directly apply the transition probability as the feature for analysis. To address the order sensitivity problem, a stacked long short-term memory network is designed to build a robust HT detection model. Following this assumption, a more sophisticated work was proposed in [52]. Lu et al. extracted information entropy instead of signal correlation from circuits as critical features [52]. This work is built based on the assumption that to maintain stealthy concealment, Trojans should be inserted in regions with low controllability and observability, which will result in low transition probability of Trojan logic.

Dong et al. [19] analyzed the existing Trojan-net features from [29] to propose five new hardware Trojan features. To further develop time-efficient approaches, they utilized the gradient boosting algorithm to train the ML model for HT detection [19]. Their algorithm applied the scoring mechanism of the eXtreme Gradient Boosting (XGBoost) [15] to set up a new feature set. Their experimental results demonstrated that they are able to obtain an average F-score of 87.75%. Similar ideas are also explored in [27] and [49], where gradient boosting algorithm was applied on features extracted from RTL level and gate level, respectively. In [72], the ML models' performance is further enhanced by hyperparameter tuning of RF model. In [76], HT detection uses class weighted XGBoost, where higher weights are assigned to minority Trojan-inserted class to remove the need for oversampling, which improves the efficiency of HT detection.

The above mentioned works focus on features that are either static signal features [18, 31] or reaction features [33]. Zhou et al. [90] analyzed the structural features of IP cores and HT triggers in gate-level circuits. Their method is specifically designed for detecting HTs triggered by less toggled signals. They abstract the circuit into a graph and extract special structures that are commonly adopted by HT implants. After obtaining the general pattern of those suspicious structures, they apply ML model for the classification work, and their experimental results yields a 100% detection rate. However, these algorithms share the same bottleneck as the above methods that require golden reference model since ML models applied here are all supervised learning. To address this limitation, Xue et al. [85] proposed a hybrid clustering model to achieve golden-Free HT detection. They achieve this by utilizing two attack models to imitate untrustworthy parties. Then through adversarial data generation, they produce the pool of adversarial samples of possible HTs, from which features of HTs were extracted. By utilizing hybrid clustering ensemble method, they were able to classify if a given circuit contains malicious implants.

Heuristic Analysis Using Unsupervised Learning Supervised learning based approaches for HT detection are promising but they require golden models. To achieve golden-free detection, unsupervised learning is utilized. The essence of unsupervised learning algorithm is to describe the distribution of data and mine potentially useful information behind, where it groups unlabeled data into different clusters based on a distance metric, as shown in Fig. 12b. First, a distance metric is defined to measure the similarity between samples. Then, clustering process happens where samples with high similarities are grouped together, so that samples outside the cluster can be distinguished. While clustering has shown promising results for HT detection [46, 73], unsupervised learning methods have two practical limitations: feature selection and distance metric. The features are commonly

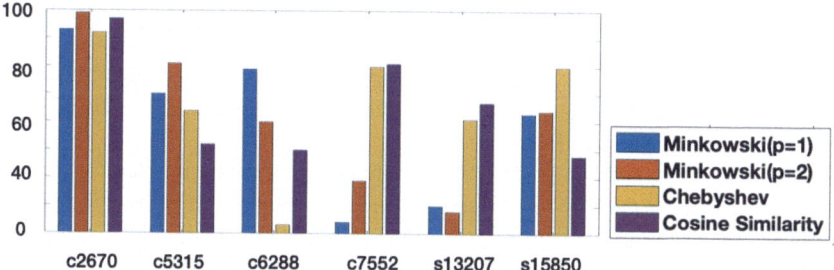

Fig. 14 The detection accuracy using clustering algorithm for several benchmark with four different distance metrics, $p = 1$ Minkowski (Manhattan), $p = 2$ Minkowski (Euclidean), Chebyshev, and Cosine similarity

selected by either expert knowledge [73] or millions of simulations [46], which may not be feasible in many scenarios. Moreover, the selection of proper distance metric is also tricky. Existing works utilize Minkowski distance as the similarity metric, which may not be suitable for HT detection. For example, Fig. 14 illustrates that clustering algorithms provide unstable performance; detection accuracy varies significantly based on the distance metric. This instability severely affects the performance of existing works. Furthermore, there are no automated methods to select a beneficial metric for a set of benchmarks with embedded HTs. Also, the performance of clustering algorithm varies from benchmark to benchmark, and measurement to measurement.

3.2 Detection of Malicious Software Attacks

This section surveys detection methods for three popular software attacks: malware attacks, ransomware attacks, and Spectre & Meltdown attacks.

3.2.1 Detection of Malware Attacks

The arms race between malware attacks and malware detection has been going on for more than two decades. In the early days, the focus of detection was on static analysis [35, 60]. The basic idea of static analysis is to utilize software filters for malware detection by extracting feature signatures by either machine learning algorithm or human expert knowledge. Unfortunately, this naive approach can be circumvented by obfuscation [2]. Dynamic detection techniques try to defend against obfuscation [7, 36]. Instead of struggling to identify static signatures despite obfuscation, such methods keep track of the runtime behavior of software and analyze any malicious behavior such as illegal access. However, both static and dynamic detection methods run on the software level. Software-based solutions, such as anti-virus software (AVS), are not effective since they rely on matching patterns that can be easily fooled by carefully crafted malware with obfuscation or other deviation capabilities. Moreover, malware can subvert AVS by abusing software vulnerabilities [37].

Researchers have recently turned their interest to hardware-based detection approaches due to their robust resistance against malware attacks compared to software-based detection. Petroni et al. [40] introduced a Peripheral Component Interconnect (PCI) based detector that monitors immutable kernel memory and successfully detects various kernel-level rootkits. Since this PCI-based method relies on physical memory address, it varies from run to run, which makes its performance unstable. Methods using Hardware Performance Counters (HPC) were proposed in [16, 41, 82], but shortcomings still exist since HPCs involve false positive rates [16] that can still be improved, along with expensive performance penalties incurred by HPC readings. PREEMPT [9] overcomes this weakness by utilizing the Embedded Trace Buffer (ETB), which gives a prediction accuracy as high as 94%. This method was refined in [69] by utilizing deep neural networks, and detection accuracy as high as 98% was obtained. Another work in [68] proposed a combinational framework that can lead to improved accuracy as well as interpretation of classification results, while addressing the HPC reliability concerns.

3.2.2 Detection of Ransomware Attacks

Ransomware attacks have been increasing throughout recent years, and many methods have been proposed to detect and prevent them. Most current ransomware detection and analysis methods fall into two main categories: static and dynamic. Dynamic analysis method involves creating an isolated environment and running the ransomware within that environment to recognize its functional behavior. Static approaches, on the other hand, include reverse engineering the malicious code to understand the working of the ransomware and then to develop defenses against it.

Sgandurra et al. proposed EldeRan [75] that checks ransomware signatures by analyzing a set of actions during the initial phases of the attack flow kill chain. EldeRan dynamically detects and classifies ransomware by analyzing operation activities such as registry key operations, calls from Windows APIs, folder and file system operations. EldeRan uses logical regression based ML model to classify each user application and has additional functionality to identify and create signatures for as yet unknown ransomware.

Andronio et al. [5] proposed the HelDroid system that detects a class of ransomware that is designed to target Android platforms. The HelDroid system uses code characteristics, including application manifests and call functions to identify ransomware and its family class using Natural Language Processing (NLP). The HelDroid system is trained to identify common messages that appear in the ransomware code to identify it. However, HelDroid's main weakness is that it uses a text classifier to detect and characterize ransomware. Mercaldo et al. [23] developed a parser that automatically identifies related ransomware instructions in a three-step process by analyzing sample code and detecting the associated Android ransomware.

3.2.3 Detection of Spectre and Meltdown Attacks

There are a large number of existing efforts for defending against Spectre and Meltdown attacks. Existing approaches focus on mitigation techniques [42, 43, 79, 86], such as enforcing the processor to empty branch target buffer during task switching, or occasionally shutting down speculative execution. However, these techniques can lead to unacceptable performance degradation. A recent approach utilizes machine learning for detection of Spectre and Meltdown attacks [3]. While it provides promising results, it has two inherent weaknesses. It cannot detect Spectre and Meltdown attacks in the presence of obfuscation techniques or other deviation capabilities [47]. Moreover, due to the black-box nature of machine learning models, the results cannot be interpreted in a meaningful way [63].

4 Summary

Security of computer systems relies on the trustworthiness of both software and hardware components. Attackers can exploit either software-level or hardware-level vulnerabilities to launch malicious attacks. There is a critical need to develop efficient defense strategies including attack detection and vulnerability mitigation mechanisms. Table 1 shows a summary of various vulnerabilities and suitable defense mechanisms. In this chapter, we first outlined the security vulnerabilities of modern computing systems from both hardware and software domains. We have also provided an overview of various attacks on machine learning models. Next, we described various detection approaches against these malicious attacks and highlighted their challenges. The subsequent chapters will describe security solutions to address these challenges.

Table 1 Vulnerabilities in computer systems and various defense mechanisms

Vulnerabilities	Attacks	Defenses
Hardware vulnerabilities	(i) Hardware Trojan	(i) Simulation-based validation
	(ii) Supply chain vulnerability	(ii) Side-channel analysis
	(iii) Reverse engineering	(iii) Heuristic analysis
	(iv) Side-channel leakage	
Software vulnerabilities	(i) Malware attack	(i) Static analysis
	(ii) Ransomware attack	(ii) Dynamic detection
	(iii) Spectre & meltdown attack	(iii) Hardware-assisted detection
		(iv) Isolated environment
Machine learning vulnerabilities	(i) Adversarial attack	(i) Model retraining
	(ii) AI Trojan attack	(ii) Pruning
		(iii) Bayesian Neural Networks

References

1. Common Weakness Enumeration. [Accessed: 08-15-2023].
2. *Malware Obfuscation Techniques: A Brief Survey*. IEEE Computer Society, 2010.
3. Bilal Ahmad. Real time detection of spectre and meltdown attacks using machine learning. *CoRR abs/2006.01442*, 2020.
4. Abdulrahman Alaql, Saranyu Chattopadhyay, Prabuddha Chakraborty, Tamzidul Hoque, and Swarup Bhunia. Lego: A learning-guided obfuscation framework for hardware IP protection. *IEEE Transactions on Computer-Aided Design of Integrated Circuits and Systems*, 2021.
5. Nicoló Andronio, Stefano Zanero, and Federico Maggi. HelDroid: Dissecting and detecting mobile ransomware. In *international symposium on recent advances in intrusion detection*, pages 382–404. Springer, 2015.
6. Giovanni Apruzzese, Michele Colajanni, Luca Ferretti, and Mirco Marchetti. Addressing adversarial attacks against security systems based on machine learning. In *2019 11th international conference on cyber conflict (CyCon)*, volume 900, pages 1–18. IEEE, 2019.
7. Daniel Arp et al. Effective and efficient malware detection at the end host. In *18th USENIX*, Montreal, Quebec, 2009.
8. Abhishek Basak, Swarup Bhunia, Thomas Tkacik, and Sandip Ray. Security assurance for system-on-chip designs with untrusted IPs. *IEEE Transactions on Information Forensics and Security*, 12(7):1515–1528, 2017.
9. Kanad Basu et al. PREEMPT: preempting malware by examining embedded processor traces. In *DAC*, page 166, 2019.
10. Swarup Bhunia, Michael S Hsiao, Mainak Banga, and Seetharam Narasimhan. Hardware Trojan attacks: Threat analysis and countermeasures. *Proceedings of the IEEE*, 102(8):1229–1247, 2014.
11. Travis Boraten and Avinash Karanth Kodi. Packet security with path sensitization for NOCs. In *2016 Design, Automation & Test in Europe Conference & Exhibition (DATE)*, pages 1136–1139. IEEE, 2016.
12. Burçin Cakır and Sharad Malik. Hardware Trojan detection for gate-level ICs using signal correlation based clustering. In *2015 Design, Automation & Test in Europe Conference & Exhibition (DATE)*, pages 471–476. IEEE, 2015.

13. R. Chakraborty et al. MERO: A statistical approach for hardware Trojan detection. In *CHES*, pages 396–410, 2009.
14. Subodha Charles and Prabhat Mishra. Reconfigurable network-on-chip security architecture. *ACM Transactions on Design Automation of Electronic Systems (TODAES)*, 25(6):1–25, 2020.
15. Tianqi Chen and Carlos Guestrin. XGBoost: A scalable tree boosting system. In *Proceedings of the 22nd ACM SIGKDD International Conference on Knowledge Discovery and Data Mining*, KDD '16, pages 785–794, New York, NY, USA, 2016. ACM.
16. John Demme et al. On the feasibility of online malware detection with performance counters. In *The 40th Annual ISCA*, pages 559–570, 2013.
17. Guneet S Dhillon, Kamyar Azizzadenesheli, Zachary C Lipton, Jeremy Bernstein, Jean Kossaifi, Aran Khanna, and Anima Anandkumar. Stochastic activation pruning for robust adversarial defense. *arXiv preprint arXiv:1803.01442*, 2018.
18. Halit Dogan, Domenic Forte, and Mark Mohammad Tehranipoor. Aging analysis for recycled FPGA detection. In *2014 IEEE international symposium on defect and fault tolerance in VLSI and nanotechnology systems (DFT)*, pages 171–176. IEEE, 2014.
19. Chen Dong, Jinghui Chen, Wenzhong Guo, and Jian Zou. A machine-learning-based hardware-Trojan detection approach for chips in the internet of things. *International Journal of Distributed Sensor Networks*, 15(12):1550147719888098, 2019.
20. Farimah Farahmandi, Yuanwen Huang, and Prabhat Mishra. *System-on-Chip Security: Validation and Verification*. Springer, 2020.
21. Nicole Fern, Ismail San, Cetin Kaya Koç, and Kwang-Ting Cheng. Hardware Trojans in incompletely specified on-chip bus systems. In *2016 Design, Automation & Test in Europe Conference & Exhibition (DATE)*, pages 527–530. IEEE, 2016.
22. Nicole Fern, Ismail San, Cetin Kaya Koç, and Kwang-Ting Tim Cheng. Hiding hardware Trojan communication channels in partially specified SoC bus functionality. *IEEE Transactions on Computer-Aided Design of Integrated Circuits and Systems*, 36(9):1435–1444, 2016.
23. Alberto Ferrante, Miroslaw Malek, Fabio Martinelli, Francesco Mercaldo, and Jelena Milosevic. Extinguishing ransomware-a hybrid approach to android ransomware detection. In *International symposium on foundations and practice of security*, pages 242–258. Springer, 2017.
24. Julien Francq and Florian Frick. Introduction to hardware Trojan detection methods. In *2015 Design, Automation & Test in Europe Conference & Exhibition (DATE)*, pages 770–775. IEEE, 2015.
25. Marc Fyrbiak, Sebastian Strauß, Christian Kison, Sebastian Wallat, Malte Elson, Nikol Rummel, and Christof Paar. Hardware reverse engineering: Overview and open challenges. In *2017 IEEE 2nd International Verification and Security Workshop (IVSW)*, pages 88–94. IEEE, 2017.
26. Syed Kamran Haider, Chenglu Jin, Masab Ahmad, Devu Manikantan Shila, Omer Khan, and Marten van Dijk. Advancing the state-of-the-art in hardware Trojans detection. *IEEE Transactions on Dependable and Secure Computing*, 16(1):18–32, 2017.
27. Tao Han, Yuze Wang, and Peng Liu. Hardware Trojans detection at register transfer level based on machine learning. In *2019 IEEE International Symposium on Circuits and Systems (ISCAS)*, pages 1–5, 2019.
28. Yi Han, Sriharsha Etigowni, Hua Liu, Saman A. Zonouz, and Athina P. Petropulu. Watch me, but don't touch me! contactless control flow monitoring via electromagnetic emanations. In *Proceedings of the 2017 ACM*, pages 1095–1108, 2017.
29. Kento Hasegawa, Masaru Oya, Masao Yanagisawa, and Nozomu Togawa. Hardware Trojans classification for gate-level netlists based on machine learning. In *2016 IEEE 22nd International Symposium on On-Line Testing and Robust System Design (IOLTS)*, pages 203–206. IEEE, 2016.
30. Kento Hasegawa, Masao Yanagisawa, and Nozomu Togawa. Hardware Trojans classification for gate-level netlists using multi-layer neural networks. In *2017 IEEE 23rd International Symposium on On-Line Testing and Robust System Design (IOLTS)*, pages 227–232. IEEE, 2017.

31. Kento Hasegawa, Masao Yanagisawa, and Nozomu Togawa. Trojan-feature extraction at gate-level netlists and its application to hardware-Trojan detection using random forest classifier. In *2017 IEEE International Symposium on Circuits and Systems (ISCAS)*, pages 1–4. IEEE, 2017.

32. Tamzidul Hoque, Jonathan Cruz, Prabuddha Chakraborty, and Swarup Bhunia. Hardware IP trust validation: Learn (the untrustworthy), and verify. In *2018 IEEE International Test Conference (ITC)*, pages 1–10. IEEE, 2018.

33. Ke Huang, John M Carulli, and Yiorgos Makris. Parametric counterfeit IC detection via support vector machines. In *2012 IEEE International Symposium on Defect and Fault Tolerance in VLSI and Nanotechnology Systems (DFT)*, pages 7–12. IEEE, 2012.

34. Yuanwen Huang, Swarup Bhunia, and Prabhat Mishra. Scalable test generation for Trojan detection using side channel analysis. *IEEE Transactions on Information Forensics and Security (TIFS)*, 13(11):2746–2760, 2018.

35. Nwokedi Idika and Aditya Mathur. A survey of malware detection techniques. *Purdue University*, 03 2007.

36. Grégoire Jacob, Hervé Debar, and Eric Filiol. Behavioral detection of malware: from a survey towards an established taxonomy. *Journal in Computer Virology*, 4(3):251–266, 2008.

37. Suman Jana and Vitaly Shmatikov. Abusing file processing in malware detectors for fun and profit. In *IEEE S&P*, 2012.

38. Yier Jin, Dzmitry Maliuk, and Yiorgos Makris. Post-deployment trust evaluation in wireless cryptographic ICs. In *2012 Design, Automation & Test in Europe Conference & Exhibition (DATE)*, pages 965–970. IEEE, 2012.

39. Yier Jin, Dzmitry Maliuk, and Yiorgos Makris. A post-deployment IC trust evaluation architecture. In *2013 IEEE 19th International On-Line Testing Symposium (IOLTS)*, pages 224–225. IEEE, 2013.

40. Nick L. Petroni Jr. et al. Copilot - a coprocessor-based kernel runtime integrity monitor. In *Proceedings of the 13th USENIX Security Symposium*, pages 179–194, 2004.

41. Mikhail Kazdagli et al. Quantifying and improving the efficiency of hardware-based mobile malware detectors. In *49th Annual IEEE/ACM MICRO*, pages 37:1–37:13, 2016.

42. Khaled Khasawneh et al. SafeSpec: Banishing the spectre of a meltdown with leakage-free speculation. In *DAC*, page 60, 2019.

43. Vladimir Kiriansky et al. DAWG: A defense against cache timing attacks in speculative execution processors. In *MICRO*, pages 974–987, 2018.

44. Paul Kocher, Jann Horn, et al. Spectre attacks: exploiting speculative execution. *Commun. ACM*, 63(7):93–101, 2020.

45. A. Kulkarni, Y. Pino, and T. Mohsenin. SVM-based real-time hardware Trojan detection for many-core platform. In *2016 17th International Symposium on Quality Electronic Design (ISQED)*, pages 362–367, 2016.

46. Amey Kulkarni, Youngok Pino, Matthew French, and Tinoosh Mohsenin. Real-time anomaly detection framework for many-core router through machine-learning techniques. *ACM Journal on Emerging Technologies in Computing Systems (JETC)*, 13(1):1–22, 2016.

47. Congmiao Li and Jean-Luc Gaudiot. Challenges in detecting an "evasive spectre". *Com. Arch. Letters*, 19(1):18–21, 2020.

48. Deqiang Li, Qianmu Li, Yanfang Ye, and Shouhuai Xu. Arms race in adversarial malware detection: A survey. *ACM Computing Surveys (CSUR)*, 55(1):1–35, 2021.

49. Konstantinos G Liakos, Georgios K Georgakilas, and Fotis C Plessas. Hardware Trojan classification at gate-level netlists based on area and power machine learning analysis. In *2021 IEEE Computer Society Annual Symposium on VLSI (ISVLSI)*, pages 412–417, 2021.

50. Yu Liu, Ke Huang, and Yiorgos Makris. Hardware Trojan detection through golden chip-free statistical side-channel fingerprinting. In *Proceedings of the 51st Annual Design Automation Conference*, pages 1–6, 2014.

51. Renjie Lu, Haihua Shen, Zhihua Feng, Huawei Li, Wei Zhao, and Xiaowei Li. HTDet: A clustering method using information entropy for hardware Trojan detection. *Tsinghua Science and Technology*, 26(1):48–61, 2021.

52. Renjie Lu, Haihua Shen, Yu Su, Huawei Li, and Xiaowei Li. GramsDet: Hardware Trojan detection based on recurrent neural network. In *2019 IEEE 28th Asian Test Symposium (ATS)*, pages 111–1115, 2019.
53. Yangdi Lyu and Prabhat Mishra. A survey of side-channel attacks on caches and countermeasures. *Journal of Hardware and Systems Security*, 2(1):33–50, 2018.
54. Yangdi Lyu and Prabhat Mishra. Efficient test generation for Trojan detection using side channel analysis. In *Design, Automation & Test in Europe Conference (DATE)*, pages 408–413, 2019.
55. Yangdi Lyu and Prabhat Mishra. Automated trigger activation by repeated maximal clique sampling. In *ASPDAC*, pages 482–487, 2020.
56. Yangdi Lyu and Prabhat Mishra. MaxSense: Side-channel sensitivity maximization for Trojan detection using statistical test patterns. *ACM Transactions on Design Automation of Electronic Systems (TODAES)*, 26(3):1–21, 2021.
57. Travis Meade, Shaojie Zhang, and Yier Jin. Netlist reverse engineering for high-level functionality reconstruction. In *2016 21st Asia and South Pacific Design Automation Conference (ASP-DAC)*, pages 655–660. IEEE, 2016.
58. Prabhat Mishra, Swarup Bhunia, and Mark Tehranipoor. *Hardware IP security and trust*. Springer, 2017.
59. Prabhat Mishra and Subodha Charles. *Network-on-chip security and privacy*. Springer, 2021.
60. A. Moser, C. Kruegel, and E. Kirda. Limits of static analysis for malware detection. pages 421–430, Dec 2007.
61. L. Moura and N. Bjørner. z3: an efficient SMT solver. In *TACAS*, pages 337–340, 2008.
62. Adib Nahiyan, Kan Xiao, Kun Yang, Yier Jin, Domenic Forte, and Mark Tehranipoor. AVFSM: A framework for identifying and mitigating vulnerabilities in FSMs. In *2016 53nd ACM/EDAC/IEEE Design Automation Conference (DAC)*, pages 1–6. IEEE, 2016.
63. Zhixin Pan and Prabhat Mishra. Automated detection of spectre and meltdown attacks using explainable machine learning. In *2021 IEEE International Symposium on Hardware Oriented Security and Trust (HOST)*, pages 24–34. IEEE, 2021.
64. Zhixin Pan and Prabhat Mishra. Automated test generation for hardware Trojan detection using reinforcement learning. page 408–413, 2021.
65. Zhixin Pan and Prabhat Mishra. Backdoor attacks on Bayesian neural networks using reverse distribution. *arXiv preprint arXiv:2205.09167*, 2022.
66. Zhixin Pan and Prabhat Mishra. A survey on hardware vulnerability analysis using machine learning. *IEEE Access*, 10:49508–49527, 2022.
67. Zhixin Pan, Jennifer Sheldon, and Prabhat Mishra. Test generation using reinforcement learning for delay-based side-channel analysis. In *IEEE/ACM International Conference On Computer Aided Design (ICCAD)*, pages 1–7, 2020.
68. Zhixin Pan, Jennifer Sheldon, and Prabhat Mishra. Hardware-assisted malware detection and localization using explainable machine learning. *IEEE Transactions on Computers*, 71(12):3308–3321, 2022.
69. Zhixin Pan, Jennifer Sheldon, Chamika Sudusinghe, Subodha Charles, and Prabhat Mishra. Hardware-assisted malware detection using machine learning. In *Design Automation and Test in Europe (DATE)*, 2021.
70. Irith Pomeranz and Sudhakar M. Reddy. A measure of quality for n-detection test sets. *IEEE Trans. Computers*, 53(11):1497–1503, 2004.
71. Devendra Rai and John Lach. Performance of delay-based Trojan detection techniques under parameter variations. In *IEEE International Workshop on Hardware-Oriented Security and Trust, HOST*, pages 58–65, 2009.
72. Matli Nishanth Reddy, MR Latchmana Kumar, Pusarla Bhaskara Sai Kumar, S Thirumalai, and M Nirmala Devi. Performance enhancement by tuning hyperparameters of random forest classifier for hardware Trojan detection. In *ICDSA*, pages 177–191. Springer, 2022.
73. Hassan Salmani. COTD: Reference-free hardware Trojan detection and recovery based on controllability and observability in gate-level netlist. *IEEE Transactions on Information Forensics and Security*, 12(2):338–350, 2016.

74. Asanka Sayakkara, Nhien-An Le-Khac, and Mark Scanlon. Facilitating electromagnetic side-channel analysis for IoT investigation: Evaluating the EMvidence framework. *Forensic Science International: Digital Investigation*, 33:301003, 2020.

75. Daniele Sgandurra, Luis Muñoz-González, Rabih Mohsen, and Emil C Lupu. Automated dynamic analysis of ransomware: Benefits, limitations and use for detection. *arXiv*, 2016.

76. Richa Sharma, Nitya Kritin Valivati, GK Sharma, and Manisha Pattanaik. A new hardware Trojan detection technique using class weighted XGBoost classifier. In *2020 24th International Symposium on VLSI Design and Test (VDAT)*, pages 1–6. IEEE, 2020.

77. Christian Szegedy, Wojciech Zaremba, Ilya Sutskever, Joan Bruna, Dumitru Erhan, Ian Goodfellow, and Rob Fergus. Intriguing properties of neural networks. *arxiv*, 12 2013.

78. Randy Torrance and Dick James. The state-of-the-art in semiconductor reverse engineering. In *Design Automation Conference*, pages 333–338, 2011.

79. Guanhua Wang et al. oo7: Low-overhead defense against spectre attacks via binary analysis. *CoRR*, abs/1807.05843, 2018.

80. Xiaoxiao Wang, Mohammad Tehranipoor, and Jim Plusquellic. Detecting malicious inclusions in secure hardware: Challenges and solutions. In *2008 IEEE International Workshop on Hardware-Oriented Security and Trust*, pages 15–19. IEEE, 2008.

81. Xinmu Wang, Yu Zheng, Abhishek Basak, and Swarup Bhunia. IIPS: Infrastructure IP for secure SoC design. *IEEE Transactions on Computers*, 64(8):2226–2238, 2014.

82. Xueyang Wang and Ramesh Karri. Numchecker: detecting kernel control-flow modifying rootkits by using hardware performance counters. In *DAC*, pages 79:1–79:7, 2013.

83. Wikipedia contributors. WannaCry ransomware attack — Wikipedia, the free encyclopedia, 2022. [Online; accessed 13-November-2022].

84. Kan Xiao, Domenic Forte, Yier Jin, Ramesh Karri, Swarup Bhunia, and Mohammad Tehranipoor. Hardware Trojans: Lessons learned after over a decade of research. *ACM Transactions on Design Automation of Electronic Systems (TODAES)*, 22(1):1–23, 2016.

85. Mingfu Xue, Rongzhen Bian, Weiqiang Liu, and Jian Wang. Defeating untrustworthy testing parties: A novel hybrid clustering ensemble based golden models-free hardware Trojan detection method. *IEEE Access*, 7:5124–5140, 2019.

86. Mengjia Yan et al. InvisiSpec: Making speculative execution invisible in the cache hierarchy. In *MICRO*, page 1076, 2019.

87. Zhixin Pan and Prabhat Mishra. Accelerating spectral normalization for enhancing robustness of deep neural networks. In *IEEE Computer Society Annual Symposium on VLSI, ISVLSI 2021, Tampa, FL, USA, July 7–9, 2021*, pages 260–265. IEEE, 2021.

88. Zhixin Pan and Prabhat Mishra. Hardware Trojan detection using Shapley ensemble boosting. pages 1127–1130, 2021.

89. Zhixin Pan and Prabhat Mishra. Td-zero: Automatic golden-free hardware Trojan detection using zero-shot learning. *IEEE Transactions on Computers*, 2022.

90. Er-Rui Zhou, Shao-Qing Li, Ji-Hua Chen, Lin Ni, Zhi-Xun Zhao, and Jun Li. A novel detection method for hardware Trojan in third party IP cores. In *2016 International Conference on Information System and Artificial Intelligence (ISAI)*, pages 528–532. IEEE, 2016.

Explainable Artificial Intelligence

1 Introduction

Artificial Intelligence (AI) enables the deployment of smart systems using Machine Learning (ML) algorithms. There is a wide variety of machine learning algorithms that are used in diverse domains, including transportation, cybersecurity, and business applications. A major problem with classical ML algorithms is that they only provide a decision without any explanation about how it reached the decision. In contrast, explainable AI provides insights into the decision-making process so that a designer can use it for various activities, such as accuracy improvement, localization of vulnerabilities, etc. In this chapter, we first provide an overview of various machine learning algorithms. Next, we discuss popular approaches for explainable machine learning.

2 Machine Learning Models

This section describes widely used machine learning models.

2.1 Support Vector Machine

Support Vector Machine (SVM) is a typical supervised learning model. Intuitively, supervised learning represents a learning process where an ML model is trained to satisfy given training samples with determined labels in advance. After the stage of training using the labeled set, the obtained ML model is expected to respond to new occurrences. SVM is trained to obtain a hyper-plane in data space to separate them out with given labels while trying to maximize the margin distance between data

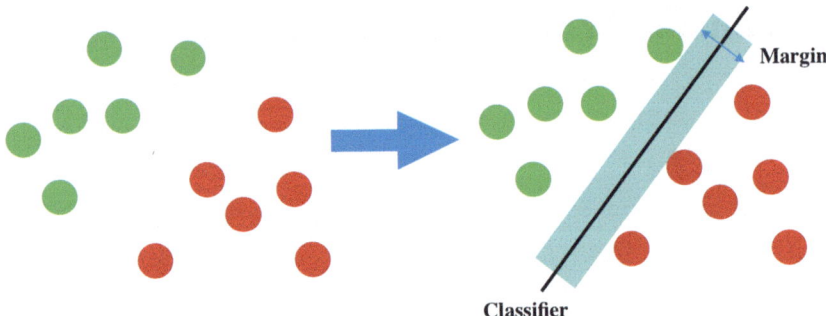

Fig. 1 A simple example of using SVM for classification

points and the hyper-plane. A trivial SVM example that achieves a bi-classification between red and green circles is shown in Fig. 1. The cyan block in the picture represents the maximized margin for an optimal classifier. Intuitively, the margin distance reflects how "far" data points from two classes are from each other, and a large margin distance indicates a better chance to correctly classify new occurrences.

2.2 Multi-Layer Perceptron

A commonly used Multi-Layer Perceptron (MLP) is a feed-forward model. An MLP contains multiple layers of neurons with an activation function, each layer fully connects the next layer with numerical values called weights. This activation function maps weighted inputs to the output of the neuron. The objective of the MLP is to learn these weights to match the inputs to the outputs as efficiently as possible. Figure 2 contains a simple example of MLP with four inputs.

2.3 Decision Tree

Decision Tree (DT) is another commonly used supervised learning algorithm for classification. An example decision tree is shown in Fig. 3 to illustrate its functionality and interpretability. The task is to decide the location for exercise based on weather conditions. As we can see, DT is a tree-structure model with nodes and edges. The basic workflow of DT is to perform a top-down tree traversal from the root. At each node, a specific attribute becomes the key factor to determine the branch. This process continues until a category label reaches the leaf node, which becomes the final decision.

The construction of such a decision tree relies on the selection of attributes for a given task. In many real-world applications, this challenge is addressed using a large

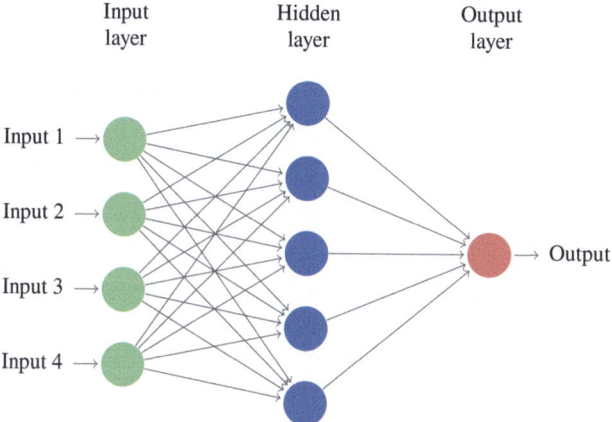

Fig. 2 A simple example of MLP with four inputs

Fig. 3 Example decision tree
to decide the exercise location
based on weather conditions

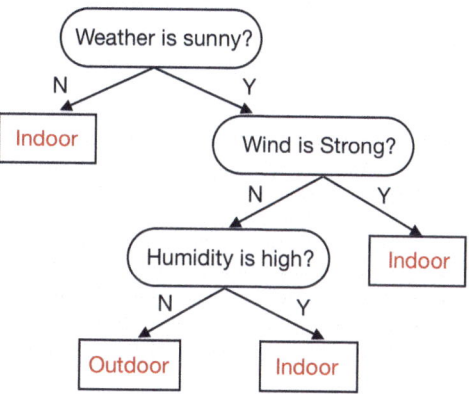

amount of historical data to automatically build the decision tree. The construction
of the tree is usually recursive and can be performed automatically.

2.4   Random Forest

Random Forest (RF), as the name suggests, is an ensemble of decision trees. RF
is actually an integrated algorithm. It first randomly selects different features and
training samples to generate many decision trees, where each decision tree is trained
with a subset of the training data. Then RF synthesizes the results of these decision
trees by voting or taking the average in the ensemble to present the final output.
Random forest is widely used in reality analysis. Compared with decision trees, it
provides significant improvement in accuracy and robustness at the same time.

2.5 Linear Regression

The motivation for Linear Regression (LR) comes from statistics. Given a dataset, LR learns from this dataset to produce a linear model that reflects the relationship between x_i and y_i as accurately as possible. Formally, the model can be written as

$$f(\mathbf{x}) = \mathbf{W}^\top \mathbf{x} + b$$

where $\mathbf{x} = \{x_1, x_2, \ldots\}$, b is the bias term, and $\mathbf{W}$ is a set of weight parameters indicating the weight of corresponding attributes. The learning process of LR is an optimization problem to obtain optimal weight parameters $\mathbf{W}$. The trained model with suitable weights is used to predict values for new inputs.

2.6 Deep Neural Network

A Deep Neural Network (DNN) is an artificial neural network that can express or simulate a wide variety of intrinsic functionalities in the fields of classification, regression, reconstruction, etc. Analogous to neurons in our nervous system, the "neural network" of artificial intelligence is a system built from "neurons" as shown in Fig. 4.

The functionality of one single neuron is limited but DNN makes use of multiple neurons and arranges them in layers. Typically, DNN consists of an input layer, an output layer, and arbitrary number of hidden layers in between for enabling DNNs to approximate the complex mapping of given data's inputs and outputs. The training process of DNNs can be summarized as follows:

1. Determine the structure and initialize the weights.
2. Feed training samples to compute training loss.

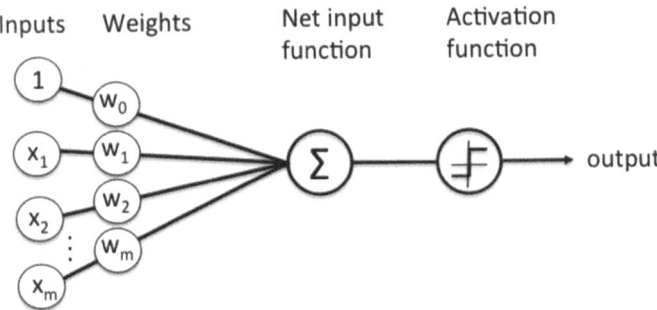

Fig. 4 A typical DNN neuron. The activation layer is applied to induce non-linearity

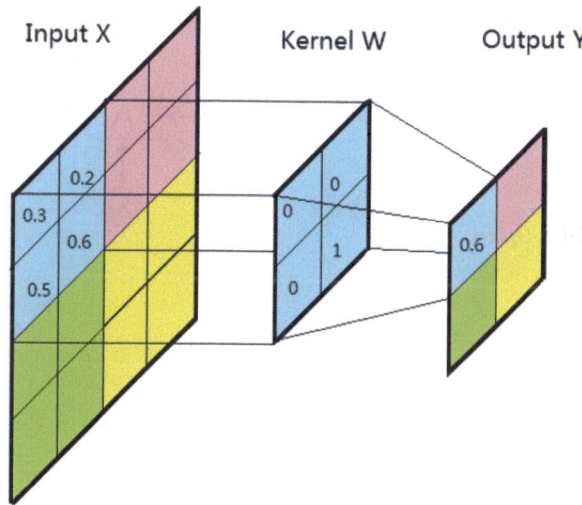

Fig. 5 Basic structure of CNN [26]

3. Compute the gradient of loss, use backpropagation [18] to update the weight coefficients, and repeat until convergence.

The flexibility of DNNs also enables their different variations to be successfully applied in various contexts. The next three ML models show three specific examples of such variations.

2.7 Convolution Neural Network

Convolution Neural Network (CNN) is a variant of DNN utilizing convolution layers. The emergence of convolutional neural networks has surpassed ordinary neural networks in the field of image processing with the characteristics of fewer parameters, fast training, high scores, and easy migration. A simple example of CNN is shown in Fig. 5. The convolution operation works by moving the convolution filters (kernel) across the input images, extracting structural features from the images at each location.

2.8 Recurrent Neural Network

Recurrent Neural Network (RNN) is another variant of neural networks. The general structure of RNN is shown in Fig. 6. In the picture, **A** represents the neural network

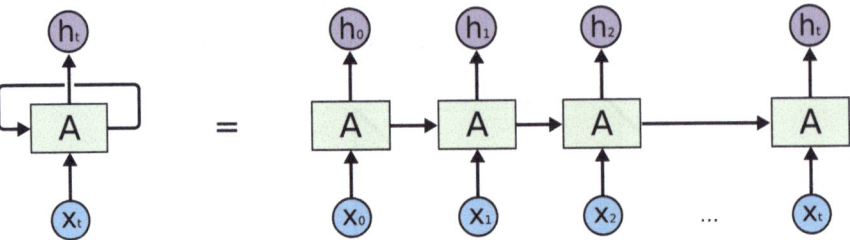

Fig. 6 Basic structure of RNN

architecture, where $x_0, x_1, x_2, \ldots, x_t$ represents the time series inputs and h_is are the outputs of hidden layers. For each single input x_i, RNN not only provides immediate response h_i but also stores the information of the current input by updating the architecture itself. Meanwhile, stored information will also be fed into the architecture in the next iteration to supply extra information. Therefore, it is widely applied in security domain as it is efficient to capture temporal dependencies of variables (signals).

2.9 Long Short-Term Memory

A specific type of RNN model is called *Long Short-Term Memory* (LSTM). It applies *gate* mechanism to solve vanishing gradient and exploding gradient. Meanwhile, the gate mechanism provides feature filtering, saving useful features and discarding useless features, which greatly enriches the information representation capacity of the model. LSTM is suitable for explainable machine learning.

2.10 Reinforcement Learning

Reinforcement learning (RL) is a branch of machine learning, but unlike the commonly known supervised learning, it is closer to human learning. Its exploration process is actually a process of gradually learning the rules of interaction through trials and responding to feedback from the environment. The RL model continuously communicates with the environment to find an optimal strategy through a series of attempts and constantly adjusts its behavior based on feedback. Figure 7 provides the basic framework of reinforcement learning.

RL framework consists of five core components: *Agent, Environment, Action, State*, and *Reward*.

Fig. 7 The basic framework
of reinforcement learning

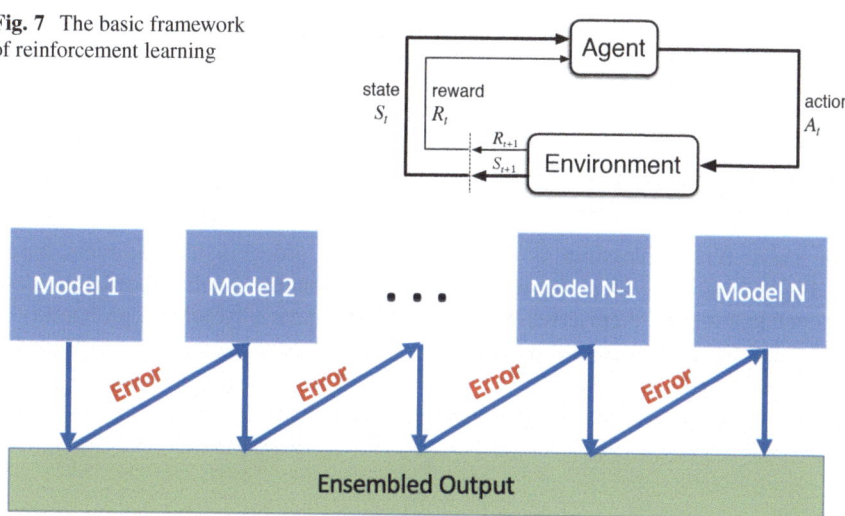

Fig. 8 The ensemble consists of a set of weak classifiers. Subsequent models focus on fixing the weakness of previous models. The final decision is based on the overall voting result

- *Agent* refers to the object that can interact with the environment through actions. The agent of reinforcement learning is usually the set of test cases to be optimized, which is continuously updated through the learning process.
- *Environment* is the receiver of the action, such as the optimization problem itself.
- *Action* consists of all possible operations that may affect the environment, such as using the current strategy for one-step calculation.
- *State* refers to information about the environment that can be perceived by the Agent, such as conditions and parameters.
- *Reward* is the feedback information from the environment that describes the effect of the latest action. For optimization problems, it often refers to the gain of objective function after performing the current operation.

2.11   Boosting

Boosting is a learning model where multiple weak learners are combined to generate a strong classifier. Figure 8 shows an overview of a boosting framework. Initially, a base weak classifier learns the training data. Next, in each iteration, a weak learner is added to reduce the training error of previously applied weak learners. Freund et al. [5] proposed the first practical Boosting algorithm, AdaBoost. The gradient boosting algorithm proposed by Breiman et al. [6] is also widely applied for many optimization problems. In each iteration of the gradient boosting algorithm, the negative gradient of the current model on all samples is calculated. Next, a new weak classifier is trained with this value for adjusting the weight of the weak classifier. Finally, the model gets updated accordingly.

2.12 Naive Bayes

The naive Bayes Classifier is one of the simple probabilistic classifiers based on Bayes' theorem. This classifier has a strong assumption that the features are independent among themselves. From the training data, a likelihood probability is calculated for each feature. For unknown input data, the posterior probability for each class is calculated using Bayes' theorem. The class having maximum posterior probability value becomes the predicted level for the input data.

2.13 Zero-Shot Learning

Zero-Shot Learning (ZSL) is a variant of transfer learning, which relies on zero training samples to handle unseen categories. The key idea of ZSL is to focus on learning the "general knowledge" of given data, and unlike the commonly known learning approaches, it is closer to the human brain when making judgments. For example, assume that a child has never seen a tiger before, and only pictures of cats and dogs are shown to the child. The child is likely to identify the tiger as a cat. The reason is that, even though the "tiger" category is never seen before, the brain is able to extract information from the picture and make comparisons with known species based on similarity. Similarly, the goal of ZSL is to train the model with clever adjustments during the training stage, so that the model is capable of exploiting information to understand unseen data.

A typical implementation of ZSL is a matching network structure as shown in Fig. 9, which consists of two major components, an extractor g_θ and a comparator f_θ. Extractor is responsible for recognizing and extracting general knowledge from training set, while compactor works by comparing the similarity of test input and known genre to assign label to it based on similarity score. In this scenario, the task is to classify the genre of dog, where the model is not trained to merely remember human-defined features from known samples but trained to be sophisticated in mining underlying features by themselves, and make reasoning by comparing afterwards. By extracting general knowledge from the entire sample set S, four different types of general knowledge were extracted by g_θ in the model. Then f_θ works by comparing the given sample's feature with pre-stored ones and computing similarity scores. The one with the highest score becomes the classified label.

This section described a wide variety of machine learning (ML) models. Table 1 provides a list of pros and cons for these ML models. This comparison will enable a designer to choose the most appropriate ML model for a given application based on the application-specific requirements.

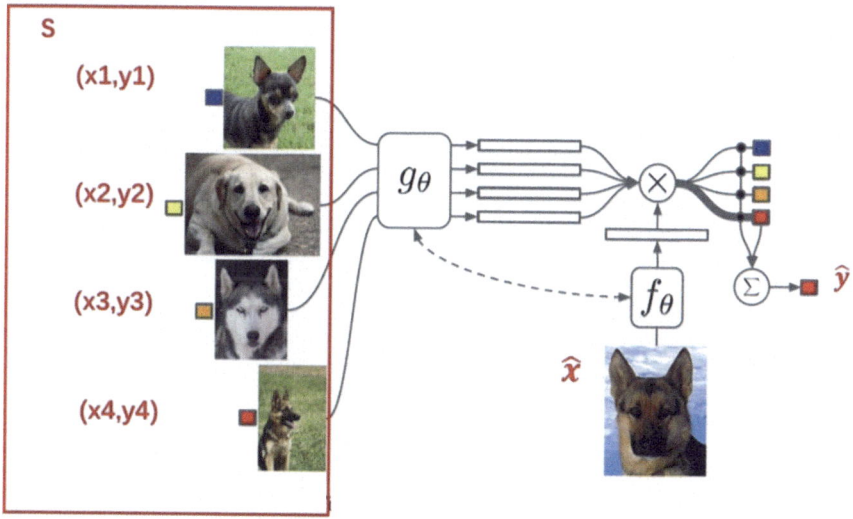

Fig. 9 ZSL utilizing matching networks as proposed in [25]

Table 1 Advantages and disadvantages of various machine learning techniques

ML Algorithms	Advantages	Disadvantages
Supervised learning (e.g., SVM, DT, RF, LR)	(i) Easy to manipulate	(i) Require golden designs or ICs
	(ii) Clear definition of decision boundary	(ii) Easy to over-fitting
	(iii) Human-understandable training process	(iii) Improper to train a large dataset
	(iv) Relatively less storage requirement	
	(v) Known number of classes prior training	
	(vi) Better transparency	
	(vii) More accurate and reliable results	
Unsupervised learning (e.g., PCA, Clustering)	(i) Need no golden designs as references	(i) Unexpected model output
	(ii) Scalable and efficient for large designs	(ii) Easy to fall into local optimization
	(iii) Good flexibility for various tasks	(iii) Sensitive to noise
		(iv) Sensitive to initialized condition

(continued)

Table 1 (continued)

ML Algorithms	Advantages	Disadvantages
Reinforcement learning	(i) Can solve very complex problems	(i) Not preferable to solve simple problems
	(ii) Models can correct errors occurred during the training process by themselves	(ii) Need huge amount of data and long training time
	(iii) Good exploration & exploitation	(iii) Based on Markovian assumption, which may not hold always
	(iv) Optimal solution where the only way for data collection is interacting with environment	(iv) Model performance heavily depends on design of reward function

3 Explainable Artificial Intelligence

In traditional machine learning frameworks, given an input sample $\mathbf{x} = \{x_1, x_2, \ldots\}$, where $x_1, x_2, \ldots$ are feature components, a classifier C will assign this instance a label $\mathbf{y}$ to indicate its prediction. However, aside from this $\mathbf{x}$ to $\mathbf{y}$ mapping, no more useful information can be gathered from this system. Also, the whole framework acts like a black box lacking transparency, which is a fundamental obstacle for users to trust the results. Explainable artificial intelligence (XAI) aims to provide an explanation as to how a model reached a decision, providing rationale beyond just the probability it uses internally to make a classification [1, 8]. The demand for explainable machine learning has been steadily increasing ever since machine learning algorithms were widely adopted in many fields, especially in security domains. Explainable AI methods can be divided into six broad categories: model interpretability, knowledge extraction, saliency maps, integrated gradients, Shapley value analysis, and layer-wise relevance propagation. The remainder of this section describes these six categories in detail.

3.1 Local Interpretability

How a model can be interpreted differs based on the model's complexity and scope. Therefore, approaches within model interpretability differentiate between local and global decisions and model-specific or model-agnostic decisions. Local interpretability explains a single step or decision a model takes in its overall decision-making process, whereas global interpretability explains all steps. Model-specific interpretability provides an explanation for a specific type of machine learning model, whereas model-agnostic interpretability provides an explanation for any type of machine learning model. One can think of model-agnostic interpretability as a generalized approach to explaining a model.

Local Interpretable Model-Agnostic Explanations (LIME) is a popular framework using model interpretability techniques [10, 14]. It aims to create an interpretable model that is *locally faithful* to the original model and its decisions. LIME performs the following three tasks.

1. LIME samples a specific instance x of the original model f to create an explanation for the original model. Once sampled, *perturbations* are generated by minor alteration of one feature of the instance and keeping the rest the same. Each perturbation is given to f to generate the real prediction for it. With enough perturbations, a local neighborhood is created around the original instance, where Π_x represents the proximity between the original instance x and the perturbations.
2. An interpretable model g is fitted on the local neighborhood. Typically, logistic regression is used to fit g, using the perturbed instances and their resulting predictions from the original model f. The measurement of local faithfulness of g to f is represented by $\mathcal{L}(f, g, \Pi_x)$.
3. The overall explanation $\xi(x)$ of the original model is calculated using the set G of interpretable models. Specifically, $\xi(x)$ is calculated by

$$\xi(x) = \operatorname*{argmin}_{g \in G} \mathcal{L}(f, g, \Pi_x) + \Omega(g) \qquad (1)$$

where $\Omega(g)$ is the complexity of the interpretable model g, as each interpretable model can vary in its resulting complexity after being fit on the local neighborhood.

Figure 10 showcases an example explanation generated by LIME for the sentiment (either positive or negative) of the sentence "This is not bad." The overall prediction resulted in the highest probability of the sentence having a negative sentiment, with LIME explaining that the biggest contributors to the negative sentiment were "bad" and "not."

Fig. 10 LIME explanation for the sentence "This is not bad" [14]

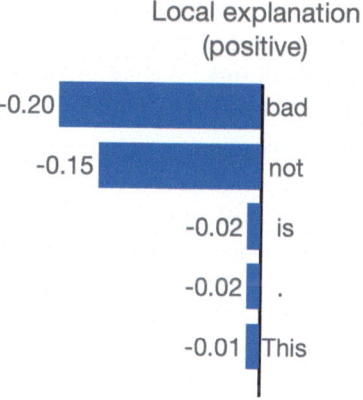

Local explanation (positive)

Even though LIME chooses the best-fitting interpretable model from G, it may still not provide a sufficient enough explanation for the original model. If the original model is very complex, it can be difficult to fully represent it with a simpler, interpretable model. The simple model may also create decisions that are not present in the original model, potentially invalidating the resulting explanation. Additionally, creating the interpretable models from the complex model is a computationally expensive task.

3.2 Knowledge Extraction

Knowledge extraction is the approach of gathering the knowledge used at the layers the model uses to make its decision. When first extracted, this knowledge will be in a form that machines can understand. Therefore, it is necessary to convert knowledge into a form that humans are able to comprehend. There are two sub-approaches within knowledge extraction: rule extraction and model distillation.

Rule Extraction Rule extraction defines a set of logical rules that exist within neural networks [3, 9, 19]. Each layer within a neural network can thus be represented by a human interpretable rule for the layer's functionality. These rules can be reconstructed into a data structure such as a decision tree like in Fig. 3 to provide an explanation for the overall model. There are two major categories of rules. First, the "if-then" rule defines a scenario where given some input x meets a condition z, the layer outputs y. Second, the "M-of-N" rules define a scenario where the layer will output y if M of N sets are satisfied based on a condition z. While these are the two major categories of rules, rules that are a subset of either category can be extracted from the original model. For example, it is possible for a model to have an "if-then-else" rule be extracted.

Rule extraction algorithms are further categorized based on locality and model-reliance [3], similar to that of the methods in Sect. 3.1. Decompositional algorithms are those that extract rules for each neuron, or decision, within the entire neural network. Pedagogical algorithms extract rules from the perspective of the entire neural network, treating the neural network as a black box. Eclectic algorithms combine the decompositional and pedagogical approaches together.

Model Distillation Model distillation extracts knowledge by compressing the decisions into human-comprehensible forms [2]. This compression of decisions is closely related to *surrogate models*, where a simple model, also known as the surrogate model, is used to represent a complex model. Model distillation aims at tackling this issue by reasonably demonstrating the reason for predicting x as y. This task is performed in three steps.

1. Select the useful features. For instance, if $\mathbf{x}$ is the feature vector $\{x_1, x_2, \ldots\}$, the user needs to sieve out the useful features while eliminating redundant ones.
2. Sort the selected features ordered by their contribution toward the final decision.

3. Analyze values of top features with the highest weights, and offer them as a human-understandable illustration. Based on the ranking status, we can provide a reasonable explanation for the behavior of classifiers.

The first step is devoted to useful feature selection. This relies on a strategy known as *Forward Propagation based Methods* [7]. This method starts with *perturbing* the inputs and then observes the changes in outputs. The perturbation can be some random noise or nullified pixels for image-based tasks. If perturbation failed to induce a relatively obvious difference in output, these features can be considered low-level contributors to the model; therefore, they can be eliminated. Conversely, if a tiny change in a feature leads to a drastic difference in prediction (output), it can be considered a major contributor.

The second step focuses on sorting the features obtained in the first step. We can directly sort the features by the magnitude of the incurred difference. However, a fundamental problem is how to accurately measure the difference. For image-based tasks, the Frobenius norm is commonly applied. For distribution-related work, KL divergence is widely adopted. However, there is no such feasible measurement for tasks in the security domain. A more stable and accurate way is using gradient analysis, also called *Backward Propagation based Methods* [21]. The backward propagation-based methods rank the importance of input features by leveraging their gradients. Both backward and forward propagation methods are built based on *white-box* setting, where we assume that users possess full access to the structure, hyperparameter, and training data. However, it may not be suitable in many scenarios due to privacy concerns or computation costs for large-scale structures. In such scenarios, *black-box* setting is required. Ribeiro et al. proposed a black-box-based algorithm [15] that starts with randomly perturbing input $\mathbf{x}$ to generate a set of artificial samples $\mathbf{x}_1, \mathbf{x}_2, \mathbf{x}_3$, etc. These samples are used by the machine learning model to obtain corresponding outputs $\mathbf{y}_1, \mathbf{y}_2, \mathbf{y}_3$, etc. Next, a linear regression leads to a linear prediction model l such that $\mathbf{y} = l(\mathbf{x})$ fits our artificial dataset. Since a linear prediction model can always be expressed by a polynomial, we can utilize the expression to extract weight information. For example, assume we have $l(\mathbf{x}) = a_1 x_1 + a_2 x_2 + a_3 x_3 + \ldots$ after computation, then we sort the terms by the absolute value of their coefficients. For example, if a_i is the largest coefficient in the term $a_i x_i$, the most important component is x_i.

The third (final) step needs to interpret these selected and sorted important features. By segmenting a given image into square sub-blocks, the explainable machine learning framework is able to sort the blocks by their contribution toward the classifier's output, so that it can illustrate what part from the given picture is able to distinguish it from other categories.

Figure 11 shows two examples of interpretation. The example on the left is from the computer vision domain where the blocks of pixels in the cat's face and ears are associated with higher coefficients, representing the primary reason for predicting a cat. The example on the right shows the table with values corresponding to different registers in different time steps. The entries associated with higher coefficients indicate the specific signals and associated clock cycles for localizing a malicious behavior.

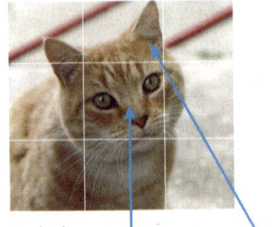

Buffer/Cycle	1	2	3	4	5	6	7	8
CS	C_1	A_1	C_3	A_3	C_5	A_5	C_7	A_7
AS	A_1	A_1	A_3	A_3	A_5	A_5	A_7	A_7
ES	E_1	B_1	B_1	E_4	B_4	B_4	E_7	B_7
DS	D_1	E_1	B_1	D_4	E_4	B_4	D_7	E_7
BS	B_1	B_1	B_1	B_4	B_4	B_4	B_7	B_7
T1	C_1	A_1	C_3	A_3	C_5	A_5	C_7	A_7
T2	D_1	E_1	B_1	D_4	E_4	B_4	D_7	E_7

$$l_1(\mathbf{x}) = a_1 x_1 + a_2 x_2 + \ldots \qquad l_2(\mathbf{x}) = b_1 x_1 + b_2 x_2 + \ldots$$

Blocks of the cat's face and ear distinguish it from other categories.

Malicious behavior most likely to happen in the sixth and first cycle.

Fig. 11 Interpret the classification results

3.3 Saliency Maps

Feature importance is an important method by which an explanation can be generated from a machine learning model. One of the simplest methods for determining the feature importance of an input to a machine learning model is saliency maps. Saliency maps are often used in the computer vision domain for images, where the gradients for the loss function are used to determine the important pixels in the image to the model's overall decision [12]. Figure 12 showcases several examples of the saliency maps for three different images using three different approaches to generate the saliency map.

The standard approach for creating the saliency map starts by computing the gradient for class probabilities based on the given input [21]. With that gradient, the score of the class given the input x can be approximated through a Taylor series. The Taylor series that approximates the score S for a class c is

$$S_c(x) \approx w^T I + b \tag{2}$$

where

$$w = \frac{\partial S_C}{\partial I}\Big|_{I_0} \tag{3}$$

is the score derivative.

While saliency maps show the most important features, it is possible that the resulting gradient is 0 for a given input, leading to saturation and the saliency map declaring the feature as not important. Smoothgrad is an alternative method to generate a saliency map to alleviate the saturation and noise [22]. First, Smoothgrad generates new images from a single baseline image by adding noise to the original image. Next, the noisy images and original image are given to the saliency map

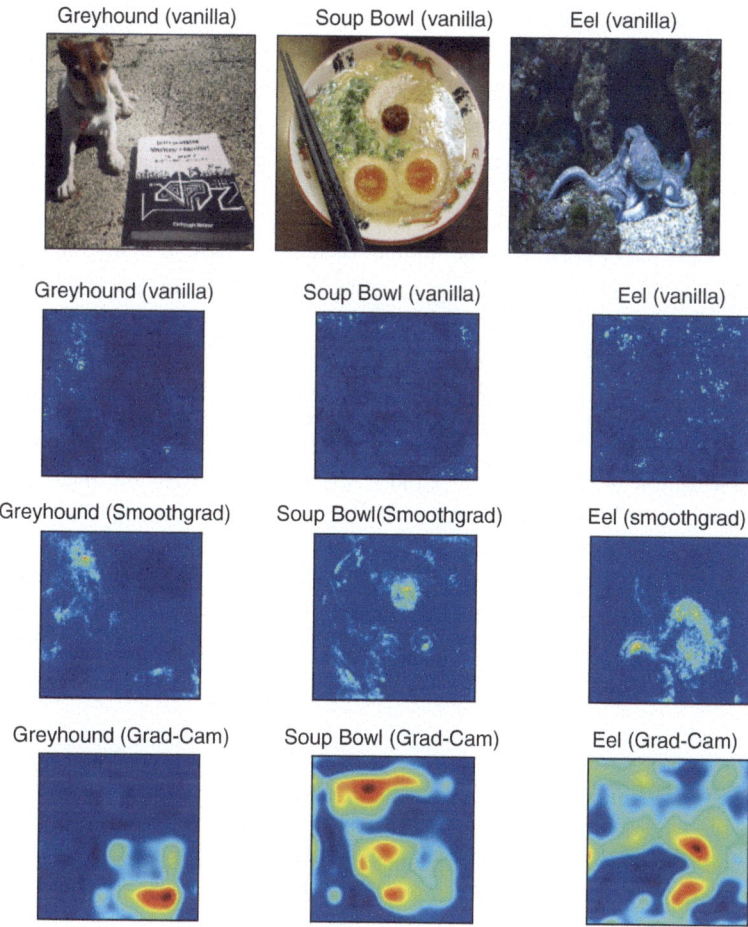

Fig. 12 Saliency map examples. Image Credit [12]

method. The returned saliency maps are then averaged to get the most important features in an image. The average map is calculated by

$$R_{smoothgrad}(x) = \frac{1}{N} \sum_{i=1}^{n} R(x + g_i) \qquad (4)$$

where g_i is a vector sampled from the Gaussian distribution. This introduces minor computational overhead compared to the standard method. The addition of noise, sampling, and averaging are all constant time and cheap operations.

Grad-CAM creates a coarse heatmap of important features through the backpropagation of gradients. Other saliency map techniques only use forward propagation. Each pixel in the resulting feature map is weighed according to

$$\alpha_k^c = \frac{1}{Z} \sum_i \sum_j \frac{\delta y^c}{\delta A_{ij}^k} \tag{5}$$

where c is the target class, k is the number of feature maps, Z is the total number of pixels, i is the width of the image, j is the height of the image, and $\frac{\delta y^c}{\delta A^k}$ is the gradient obtained via backpropagation.

Even with more optimized techniques like Smoothgrad and Grad-CAM, saliency maps do not compare the result to ground truth. This can lead to areas being highlighted in the saliency map as important, which is far from reality. For example, in Fig. 12, all three methods place importance on the book in the image for the classification of a greyhound. Grad-CAM is the worst of the three, placing importance on the book only. Smoothgrad does the best, placing importance mostly on the greyhound's face. While Grad-CAM provides inferior performance for the greyhound image, it performs well compared to the other two methods for the soup bowl. This leads to the conclusion that human oversight is necessary to ensure that the important features line up with what a human would reasonably conclude as being important.

3.4 Integrated Gradients

Integrated gradients extend the gradients calculated in a machine learning model to determine the importance of a feature to the overall decision [24]. It does so by integrating the gradient for a specific instance of the model and its resulting prediction along the path between the baseline input x' and the input x. Formally, the integrated gradient along the ith dimension IG_i is calculated as

$$IG_i(x) = (x_i - x_i') \times \int_{\alpha=0}^1 \frac{\partial F(x' + \alpha \times (x - x'))}{\partial x_i} \, d\alpha \tag{6}$$

where $\frac{\partial F(x)}{\partial x_i}$ is the ith dimension gradient of $F(x)$. Intuitively, IG_i calculates the integrated gradient for a specific feature and input through the path between the input and the baseline input and the average gradient of the feature along that path.

To produce good explanations, integrated gradients fulfill two properties: *sensitivity* and *implementation invariance*. Sensitivity means that the baseline and the given input only differ in value in one feature for each attribution, while the remaining features remain the same. Implementation invariance means that the attributions remain the same for models that are functionally equivalent.

Original image   Integrated gradients   Gradients at image

Fig. 13 Integrated gradients vs. normal gradients. Left image is original, middle image is the result from applying integrated gradients, and the right image is the result from applying normal gradients. Image credit [24]

When compared with the non-integrated gradient values, integrated gradients have a much clearer definition of the important features. Figure 13 shows the difference between the important pixels as calculated by integrated gradients and normal gradients [24]. In each example image, the integrated gradient results have much higher fidelity and clarity as to the most important pixels. Looking at the last row of images, a classifier is able to predict that the original image is of a school bus. The most important pixels in the image that contributed to this model's decision are the pixels of the "SCHOOL BUS" sign located above the windshield. While both integrated gradients and normal gradients highlight this area as important, integrated gradients provide a much clearer picture in comparison to the normal gradients.

One drawback of Eq. 6 is that the path between the input and baseline must be straight. However, not all models will have a straight path. Therefore, the integrated gradient can be calculated by representing the path as a function γ, as follows:

$$PathIG_i^\gamma(x) = \int_{\alpha=0}^{1} \frac{\partial(\gamma(\alpha))}{\partial\gamma_i(\alpha)} \frac{\partial\gamma_i(\alpha)}{\partial\alpha} \, d\alpha \qquad (7)$$

Integrated gradients introduce minor computational complexity to the original model. The original model already computes the gradients, and with integrated gradients, the model only needs to save the gradients, slightly increasing the storage requirements. After saving the gradients, integrated gradients only need to integrate them to get the important features.

Computing the integrated gradient can be sped up by approximately computing it by replacing the integral with a Riemann approximation. The approximated integrated gradient is

$$IG_i^{approx}(x) = (x_i - x_i') \times \sum_{k=1}^{m} \frac{\partial F(x' + \frac{k}{m} \times (x - x'))}{\partial x_i} \times \frac{1}{m} \tag{8}$$

where m is the number of Riemann approximation steps. Choosing a good value for m will lead to a strong approximation of the integral without needing complicated mathematical steps. Having a summation instead of an integral also enables a simpler implementation that is less likely to miscalculate. $PathIG$ can also be approximated in a similar manner using the Riemann approximation in place of the integral. This method has been extended to combine with other approaches, such as saliency maps.

3.5 Shapley Value Analysis

The concept of Shapley values (SHAP) is borrowed from cooperative game theory [16, 20]. It is used to fairly attribute a player's contribution to the end result of a game. SHAP captures the marginal contribution of each player to the final result. Formally, we can calculate the marginal contribution of the i-th player in the game by

$$\phi_i = \sum_{S \subseteq N/\{i\}} \frac{|S|!(M - |S| - 1)!}{M!} [f_x(S \cup \{i\}) - f_x(S)] \tag{9}$$

where the total number of players is $|M|$, S represents any subset of players that does not include the i-th player, and $f_x(\cdot)$ represents the function to give the game result for the subset S. Intuitively, SHAP is a weighted average payoff gain that player i provides if added into every possible coalition without i. Table 2 shows an illustrative example of computing the Shapley value of the first feature for a ML model with three features. The loss function ($\mathcal{L}$) serves as the "score" function to indicate how much payoff currently we have by applying existing features. For example, in the first row, the sequence is $1, 2, 3$, meaning we sequentially add the first, second, and third features into consideration for classification. Here, $\emptyset$ stands for the model without considering any features, which in our case is a random guess classifier and $\mathcal{L}(\emptyset)$ is the corresponding loss. Then by adding the first feature into the scenario, we use $\{1\}$ to represent the dummy model that only uses this feature to

Table 2 Contribution of first feature

Sequences	Marginal contributions
1,2,3	$\mathcal{L}(\{1\}) - \mathcal{L}(\emptyset)$
1,3,2	$\mathcal{L}(\{1\}) - \mathcal{L}(\emptyset)$
2,1,3	$\mathcal{L}(\{1, 2\}) - \mathcal{L}(\{2\})$
2,3,1	$\mathcal{L}(\{1, 2, 3\}) - \mathcal{L}(\{2, 3\})$
3,1,2	$\mathcal{L}(\{1, 3\}) - \mathcal{L}(\{3\})$
3,2,1	$\mathcal{L}(\{1, 2, 3\}) - \mathcal{L}(\{3, 2\})$

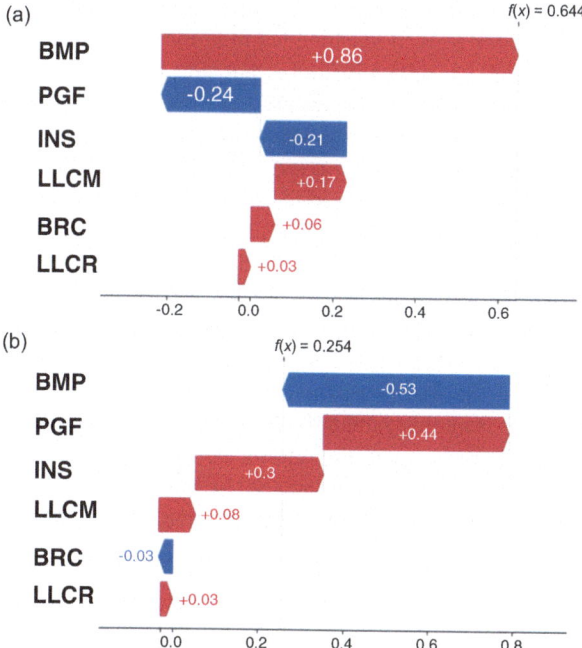

Fig. 14 Shapley values for classifying (**a**) Spectre attack and (**b**) Meltdown attack using six features: branch misprediction rate (BMP), number of page faults (PGF), number of instructions (INS), number of low-level cache misses (LLCM), number of branch instructions (BRC), and number of low-level cache references (LLCR)

perform prediction. We obtain the Shapley value for the first feature by computing the marginal contributions of all six sequences and taking the average.

The prediction of the model is the sum of the SHAP values, defined as

$$f(x^*) = f_x(\emptyset) + \sum_i \phi(i) \tag{10}$$

Figure 14 shows the Shapley values for a classifier that classifies six different features of a program and whether it is vulnerable to a Spectre or Meltdown attack [13]. Positive values indicate that the feature is supporting the input toward the positive prediction, whereas negative values indicate that the feature is supporting the input toward the negative prediction. Therefore, features with the greatest positive value indicate that they contribute to a sample being classified as an attack the most.

Shapley values can provide more insight into the explanation than other explainable methods. They provide insights into the individual features with their specific contributions to the overall result. They are also able to take into account related features that interact with each other throughout the model.

However, as SHAP is model-agnostic, it requires more computational complexity to get an explanation for a given machine learning model when compared to other model-specific methods, specifically those that are gradient-based like saliency maps. Additionally, if the Shapley values are calculated using sparse training data, it is possible that some of the attribute values will be inaccurate. For example, if there is a feature that the model did not use to make its predictions yet remains in the data, SHAP may attribute the unused feature with some importance [23].

To alleviate some of the computational complexity, SHAP can be approximated using linear regression or Monte Carlo permutation sampling [11, 17]. The linear regression approximation is done by solving the weighted least squares problem with subsampled coalitions. However, it is possible this approximation method can potentially bias the results and place more emphasis on a feature that does not actually contribute to the model and its decision-making process. Monte Carlo permutation sampling samples randomly from the player permutations and then determines the difference between the current permutation and the previous one. After comparing, the new permutations and SHAP values from it are added into the previous permutations and values. This continues until all permutations have been evaluated.

3.6 Layer-Wise Relevance Propagation

Layer-wise Relevance Propagation (LRP) is a model-specific explainable technique that uses backpropagation to create an explanation for a prediction [4]. Starting at the output layer of the model, the prediction $f(x)$ for a given input x has a relevance R denoting the explanation. R is then redistributed to the previous layer, and it continues until it reaches the input layer, where it indicates the contribution of each original input feature in the prediction. Formally, this is defined as

$$f(x) = \sum_k R_k^{l+1} = \sum_j R_j^l = \cdots = \sum_i R_i^0 \tag{11}$$

where the relevance for the ith input is R_i^0 and the relevance for the jth neuron at layer l is R_j^l.

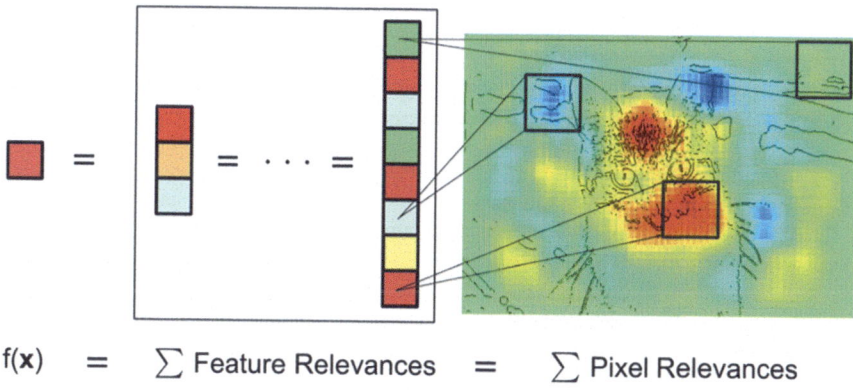

f(x) = $\sum$ Feature Relevances = $\sum$ Pixel Relevances

Fig. 15 Redistributing relevance from output layer to input layer. Image credit [4]

Figure 15 shows an example of LRP for an image of a cat. Starting at the output $f(x)$ layer, the feature relevances are then summed for each layer, iteratively. When it reaches the input layer, the pixels with the most importance in predicting $f(x)$ are highlighted.

Despite LRP being model-specific, it has been extended to other models besides CNNs that it was originally based on. LRP is computationally inexpensive as a single backward pass is needed to get the explanation for the current input and resulting prediction for a given model. Additionally, an explanation can be given from a specific starting layer instead of the input layer by stopping the backward propagation from proceeding. However, care needs to be taken to ensure that the relevance is correctly carried out through each layer. It is possible to cause an error in calculating the feature relevance at a layer; an error which will be carried out throughout the subsequent relevance calculations.

4 Summary

This chapter first introduced a wide variety of machine learning (ML) models. Next, it described six popular explainable AI methods. Table 3 compares advantages and disadvantages of these explainable AI methods so that a designer can choose the most appropriate method for a given application based on the application-specific requirements.

Table 3 Advantages and disadvantages of various explainable AI methods

XAI method	Advantages	Disadvantages
Local interpretability	(i) Local and global decisions	(i) Computationally expensive
	(ii) Can be model-specific or model-agnostic	(ii) Fitting simple models from complex models can cause errors
Knowledge extraction	(i) Explanations in human-understandable form	(i) Can be computationally expensive
Saliency maps	(i) Simple approach	(i) Only useful for images
	(ii) Small overhead, only need to track gradients	(ii) Can be affected by gradient saturation
		(iii) Without human oversight, no ground truth to compare results
Integrated gradients	(i) Better than normal gradients	(i) Requires a baseline to compare
	(ii) Can approximate results for faster computation	(ii) Model must be differentiable
	(iii) Small overhead, only need to track gradients	
Shapley value analysis	(i) Model-agnostic	(i) Computational complexity
	(ii) Decomposable explanation	(ii) Potential feature obfuscation
Layer-wise Relevance Propagation	(i) Uses internal structure of model	(i) Does not work for all architectures
	(ii) Can be extended to most model types	(ii) Layers can introduce errors during the relevance calculation
	(iii) Computational complexity	

References

1. Amina Adadi and Mohammed Berrada.   Peeking Inside the Black-Box: A Survey on Explainable Artificial Intelligence (XAI). *IEEE Access*, 6:52138–52160, 2018.
2. Raed Alharbi, Minh N. Vu, and My T. Thai.   Learning Interpretation with Explainable Knowledge Distillation, November 2021.
3. Robert Andrews, Joachim Diederich, and Alan B. Tickle. Survey and critique of techniques for extracting rules from trained artificial neural networks. *Knowledge-Based Systems*, 8(6):373–389, December 1995.
4. Sebastian Bach, Alexander Binder, Grégoire Montavon, Frederick Klauschen, Klaus-Robert Müller, and Wojciech Samek. On Pixel-Wise Explanations for Non-Linear Classifier Decisions by Layer-Wise Relevance Propagation. *PLOS ONE*, 10(7):e0130140, July 2015.
5. Yoav Freund and Robert E Schapire.   A Decision-Theoretic Generalization of On-Line Learning and an Application to Boosting.   *Journal of Computer and System Sciences*, 55(1):119–139, August 1997.
6. Jerome H. Friedman. Greedy Function Approximation: A Gradient Boosting Machine. *The Annals of Statistics*, 29(5):1189–1232, 2001.

7. Timon Gehr, Matthew Mirman, Dana Drachsler-Cohen, Petar Tsankov, Swarat Chaudhuri, and Martin Vechev. AI2: Safety and Robustness Certification of Neural Networks with Abstract Interpretation. In *2018 IEEE Symposium on Security and Privacy (SP)*, pages 3–18, May 2018.
8. David Gunning. Explainable Artificial Intelligence (XAI). 2017.
9. Tameru Hailesilassie. Rule Extraction Algorithm for Deep Neural Networks: A Review, September 2016.
10. Andreas Holzinger, Anna Saranti, Christoph Molnar, Przemyslaw Biecek, and Wojciech Samek. Explainable AI Methods - A Brief Overview. In Andreas Holzinger, Randy Goebel, Ruth Fong, Taesup Moon, Klaus-Robert Müller, and Wojciech Samek, editors, *xxAI - Beyond Explainable AI: International Workshop, Held in Conjunction with ICML 2020, July 18, 2020, Vienna, Austria, Revised and Extended Papers*, Lecture Notes in Computer Science, pages 13–38. Springer International Publishing, 2022.
11. Scott M Lundberg and Su-In Lee. A Unified Approach to Interpreting Model Predictions. In *Advances in Neural Information Processing Systems*, volume 30. Curran Associates, Inc., 2017.
12. Christoph Molnar. *Interpretable Machine Learning. A Guide for Making Black Box Models Explainable*. 2 edition, 2022.
13. Zhixin Pan and Prabhat Mishra. Automated detection of spectre and meltdown attacks using explainable machine learning. In *2021 IEEE International Symposium on Hardware Oriented Security and Trust (HOST)*, pages 24–34. IEEE, 2021.
14. Marco Tulio Ribeiro, Sameer Singh, and Carlos Guestrin. Model-Agnostic Interpretability of Machine Learning, June 2016.
15. Marco Tulio Ribeiro, Sameer Singh, and Carlos Guestrin. "Why Should I Trust You?": Explaining the Predictions of Any Classifier. In *Proceedings of the 22nd ACM SIGKDD International Conference on Knowledge Discovery and Data Mining*, KDD '16, pages 1135–1144, New York, NY, USA, August 2016. Association for Computing Machinery.
16. Alvin E Roth. *The Shapley Value: Essays in Honor of Lloyd S. Shapley*. Cambridge University Press, 1988.
17. Benedek Rozemberczki, Lauren Watson, Péter Bayer, Hao-Tsung Yang, Olivér Kiss, Sebastian Nilsson, and Rik Sarkar. The Shapley Value in Machine Learning, May 2022.
18. Sebastian Ruder. An overview of gradient descent optimization algorithms, 2016.
19. Rudy Setiono and Huan Liu. Understanding Neural Networks via Rule Extraction. In *IJCAI*, volume 1, pages 480–485, 1995.
20. Lloyd S Shapley et al. A value for n-person games. 1953.
21. Karen Simonyan, Andrea Vedaldi, and Andrew Zisserman. Deep Inside Convolutional Networks: Visualising Image Classification Models and Saliency Maps, April 2013.
22. Daniel Smilkov, Nikhil Thorat, Been Kim, Fernanda Viégas, and Martin Wattenberg. Smooth-Grad: Removing noise by adding noise, June 2017.
23. Mukund Sundararajan and Amir Najmi. The Many Shapley Values for Model Explanation. In *Proceedings of the 37th International Conference on Machine Learning*, pages 9269–9278. PMLR, November 2020.
24. Mukund Sundararajan, Ankur Taly, and Qiqi Yan. Axiomatic Attribution for Deep Networks. In *Proceedings of the 34th International Conference on Machine Learning*, pages 3319–3328. PMLR, July 2017.
25. Oriol Vinyals, Charles Blundell, Timothy Lillicrap, Koray Kavukcuoglu, and Daan Wierstra. Matching Networks for One Shot Learning. In *Advances in Neural Information Processing Systems*, volume 29. Curran Associates, Inc., 2016.
26. Zhixin Pan and Prabhat Mishra. Accelerating spectral normalization for enhancing robustness of deep neural networks. In *IEEE Computer Society Annual Symposium on VLSI, ISVLSI 2021, Tampa, FL, USA, July 7–9, 2021*, pages 260–265. IEEE, 2021.

Part II
Detection of Software Vulnerabilities

Malware Detection Using Explainable AI

1 Introduction

Malicious software (malware) is any software designed to harm a computer, server, or computer network and cause severe damage to the target system. The portability of malware also enables them to proliferate across various platforms at an alarming rate. Especially with the rapid development of Internet and smart phones in recent years, malware-implanted applications provided by third-party developers expose embedded systems to a great threat of privacy leakage due to their open-source nature. Figure 1 shows the results of a cybercrime study involving 355 companies across 11 countries covering 16 industrial sectors. It highlights that malware is the most expensive attack for organizations, with an average revenue loss of $2.6 million per organization in 2018 (11% increase compared to 2017) [6]. Clearly, there is an urgent need to develop efficient malware detection techniques.

Malware detection is a "cat and mouse" game where researchers design novel methods for malware detection, and attackers develop devious ways to circumvent detection. Signature-based detection is one of the most popular commercial malware detection techniques [5]. Signature-based detectors compare the signature of a program executable with previously stored malware signatures. However, signature-based anti-virus software (AVS) is not useful for unknown zero-delay malware since the respective signature is absent from the database. In fact, signature-based AVS is not effective even for known malware with polymorphic or metamorphic features. These morphic malware have either a mutation engine or rewrite themselves in each iteration through various program obfuscation techniques. While behavior-based AVS is promising in detecting unknown and morphing malware, they are computation-intensive. As a result, they are not suitable for resource-constrained systems such as IoT edge devices that operate under real-time, power, and energy constraints.

© The Author(s), under exclusive license to Springer Nature Switzerland AG 2023
Z. Pan, P. Mishra, *Explainable AI for Cybersecurity*,
https://doi.org/10.1007/978-3-031-46479-9_3

	Business disruption		Information loss		Revenue loss		Equipment damage		Total cost by attack type	
Malware (+11%)	$	0.5	$	1.4	$	0.6	$	0.1	$	2.6
Web-based attacks (+17%)	$	0.3	$	1.4	$	0.6	$	–	$	2.3
Denial-of-service (+10%)	$	1.1	$	0.2	$	0.4	$	0.1	$	1.7
Malicious insiders (+15%)	$	0.6	$	0.6	$	0.3	$	0.1	$	1.6
Phishing and social engineering (+8%)	$	0.4	$	0.7	$	0.3	$	–	$	1.4
Malicious code (+9%)	$	0.2	$	0.9	$	0.2	$	–	$	1.4
Stolen devices (+12%)	$	0.4	$	0.4	$	0.1	$	0.1	$	1.0
Ransomware (+21%)	$	0.2	$	0.3	$	0.1	$	0.1	$	0.7
Botnets (+12%)	$	0.1	$	0.2	$	0.1	$	–	$	0.4
Total cost by consequence	$	4.0	$	5.9	$	2.6	$	0.5	$	**13.0**

Fig. 1 Consequences of different types of cyber-attacks in 2018. The average cost of malware attacks is $2.6 million, which is an 11% increase compared to 2017 [6]

Recent research efforts explored designing hardware-assisted malware detection with the hardware as a root of trust [21]. The underlying assumption is that although AVS can be fooled by variations in malware code, it is difficult to subvert a hardware-based detector since the malware functionality will remain the same. There are some promising directions for hardware-assisted malware detection using embedded trace buffer (ETB) and hardware performance counters (HPCs). ETB-based malware detection [4] shows advantages over HPC-based methods [29] in terms of classification accuracy. Despite all these advantages, exploiting hardware components for malware detection is still in its infancy—there is no strong theoretical basis. Machine learning [18] has been successfully used for malware detection [3, 8, 11, 20, 26, 28, 31]. However, none of the previous works on machine learning-based malware detection are explainable. Therefore, the detection results cannot be interpreted in a meaningful way.

In this chapter, we introduce a hardware-assisted malware detection that takes advantage of explainable machine learning. Specifically, we discuss a hardware-assisted malware detection framework using model distillation, which leads to interpretable detection results as well as improved accuracy compared to the state-of-the-art methods. The interpretation of results sheds light on why the classifier makes incorrect decisions. This information leads to malware localization.

2 Background and Related Work

In this section, we first survey related efforts on malware detection to understand the challenges. Next, we briefly describe explainable machine learning to highlight its suitability for malware detection. Chapter "Explainable Artificial Intelligence" describes explainable AI in detail.

2.1 Malware Detection Challenges

The arms race between malware attacks and malware detection has been going on for more than two decades. In the early days, the focus of detection was on a static analysis [13, 19]. The basic idea of static analysis is to utilize software filters for malware detection by extracting feature signatures by either machine learning algorithm or human expert knowledge. Unfortunately, this naive approach can be circumvented by obfuscation [1]. Dynamic detection techniques try to defend against obfuscation [3, 14]. Instead of struggling with concealing strings created by obfuscation, such methods keep track of the runtime behavior of software and analyze any malicious behavior such as illegal access. However, both static and dynamic detection methods run on the software level. AVS is unable to detect malware with obfuscation or other deviation capabilities. Moreover, malware can subvert AVS by abusing software vulnerabilities [15].

Recent research efforts turned interest into hardware-based detection approaches due to their robust resistance against malware attacks compared to software-based detection. Petroni et al. [16] introduced a Peripheral Component Interconnect (PCI)-based detector that monitors immutable kernel memory and successfully detects various kernel-level rootkits. Since this PCI-based method relies on physical memory address, it varies from run to run, which makes its performance unstable. Methods using Hardware Performance Counters (HPCs) were proposed in [9, 17, 30], but the shortcoming still exists since HPCs involve unacceptably high false positive rates [9], along with expensive performance penalties incurred by HPC readings. PREEMPT [4] overcomes this weakness by utilizing the Embedded Trace Buffer (ETB), which gives a prediction accuracy as high as 94%. While hardware-based prediction is promising, it inherits three fundamental limitations. (1) These detection methods make prediction based on features collected from single cycles separately without considering the interaction between consecutive cycles. For malicious behaviors activated in consecutive time slots, it is hard to gather sufficient information. (2) Since the execution of malware consists of both normal (benign) execution and malicious computation, the existing methods require expensive preprocessing to eliminate useless benign cycles. (3) Most importantly, a user gets only the final decision without understanding how the decision was made or where to locate the infected area. This can also lead to inaccurate predictions due to various factors including incorrect tracing, improper preprocessing, or erroneous machine learning algorithms.

One promising avenue to address these challenges is to utilize explainable AI. First, it interprets the outputs of a machine learning model with a ranking of contribution factors, which explicitly provide a detailed feature importance map and explain the internal mechanism of each individual prediction. Moreover, we adopt the idea of long short-time memory that grants our model the ability to handle time-sequential data, which is crucial in collaborating with real-time hardware components.

2.2 Why Explainable AI for Malware Detection?

The demand for explainable machine learning has been steadily increasing ever since machine learning algorithms were widely adopted in many fields, especially in security domains. Specifically, in traditional machine learning frameworks, given an input sample $\mathbf{x} = \{x_1, x_2, \ldots\}$, where $x_1, x_2, \ldots$ are feature components, a classifier C will assign this instance a label $\mathbf{y}$ to indicate its prediction. However, aside from this $\mathbf{x}$ to $\mathbf{y}$ mapping, no more useful information can be gathered from this system. Also, the whole framework acts like a black-box lacking transparency, which is a fundamental obstacle for users to trust the results.

Explanation schemes in machine learning aim at tackling this issue by reasonably demonstrating the reason for predicting x as y. This task is performed in three steps. (1) Select the useful features. For instance, in case $\mathbf{x}$ is the feature vector $\{x_1, x_2, \ldots\}$, the user needs to sieve out the useful features while eliminating redundant ones. (2) Sort the selected features ordered by their contribution toward the final decision. (3) Analyze values of top features with highest weights, and offer them as a human-understandable illustration. Based on the ranking status, we can provide a reasonable explanation for the behavior of classifiers.

The first step is devoted to useful feature selection. This relies on a strategy known as *Forward Propagation-Based Methods* [10]. This method starts with *perturbing* the inputs and then observes the changes in outputs. The perturbation can be some random noise or nullify pixels for image-based tasks. If perturbation failed to induce relatively obvious difference in output, these features can be considered as low-level contributors to the model, and therefore, they can be eliminated. Conversely, if a tiny change of a feature leads to a drastic difference in prediction (output), it can be considered as a major contributor.

The second step focuses on sorting the features obtained in the first step. We can directly sort the features by the magnitude of the incurred difference. However, a fundamental problem is how to accurately measure the difference. For image-based tasks, the Frobenius norm is commonly applied. For distribution-related work, KL divergence is widely adopted. However, there is no such feasible measurement for tasks in security domain. A more stable and accurate way is using gradient analysis, also called *Backward Propagation-Based Methods* [27]. The backward propagation-based methods rank the importance of input features by leveraging their gradients.

Both backward and forward propagation methods are built based on *white-box* setting, where we assume that users possess full access to the structure, hyper parameter, and training data. However, it may not be suitable in many scenarios due to the privacy concern or computation cost for large-scale structures. In such scenarios, *black-box* setting is required. Ribeiro et al. proposed a black-box-based algorithm [23] that starts with randomly perturbing input $\mathbf{x}$ to generate a set of artificial samples $\mathbf{x}_1, \mathbf{x}_2, \mathbf{x}_3$, etc. These samples are used by the machine learning model to obtain corresponding outputs $\mathbf{y}_1, \mathbf{y}_2, \mathbf{y}_3$, etc. Next, a linear regression

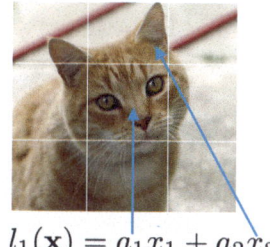

Buffer/Cycle	1	2	3	4	5	6	7	8
CS	C_1	A_1	C_3	A_3	C_5	A_5	C_7	A_7
AS	A_1	A_1	A_3	A_3	A_5	A_5	A_7	A_7
ES	E_1	B_1	B_1	E_4	B_4	B_4	E_7	B_7
DS	D_1	E_1	B_1	D_4	E_4	B_4	D_7	E_7
BS	B_1	B_1	B_1	B_4	B_4	B_4	B_7	B_7
TI	C_1	A_1	C_3	A_3	C_5	A_5	C_7	A_7
T2	D_1	E_1	B_1	D_4	E_4	B_4	D_7	E_7

$$l_1(\mathbf{x}) = a_1 x_1 + a_2 x_2 + \ldots$$

$$l_2(\mathbf{x}) = b_1 x_1 + b_2 x_2 + \ldots$$

Blocks of the cat's face and ear distinguish it from other categories.

Malicious behavior most likely to happen in the sixth and first cycle.

Fig. 2 Interpretation of the classification results

leads to a linear prediction model l such that $\mathbf{y} = l(\mathbf{x})$ fits our artificial dataset. Since a linear prediction model can always be expressed by a polynomial, we can utilize the expression to extract weight information. For example, assume we have $l(\mathbf{x}) = a_1 x_1 + a_2 x_2 + a_3 x_3 + \ldots$ after computation, and then we sort the terms by absolute value of their coefficients. For example, if a_i is the largest coefficient in the term $a_i x_i$, the most important component is x_i.

The third (final) step needs to interpret these selected and sorted important features. A simple example in the computer vision domain is image recognition as shown in Fig. 2. By segmenting the given image into square sub-blocks, the explainable machine learning framework is able to sort the blocks by their contribution toward the classifier's output, so that it can illustrate what part from the given picture is able to distinguish it from other categories.

While the existing approaches are promising, they face two fundamental challenges in dealing with hardware-assisted malware detection. (1) Existing approaches consider input data that are static pixel images. However, in malware detection, we need to handle input data that are time-sequential records. (2) While linear regression can be used by existing methods, it can lead to serious accuracy loss when dealing with real datasets since linear regression suffers from the sensitivity to extreme values like isolation points. Even piecewise linear regression may not work due to high computation cost.

3 Malware Detection Using Explainable Machine Learning

To address the challenges of malware detection, we need to develop a framework that enables the synergistic integration of both hardware trace analysis and explainable machine learning for efficient malware detection. We utilize existing design-for-debug architecture, such as embedded trace buffer, for trace collection. Such traces can be viewed as a $w \times d$ table $\mathbf{X}$, where w is the *width*, and d is the

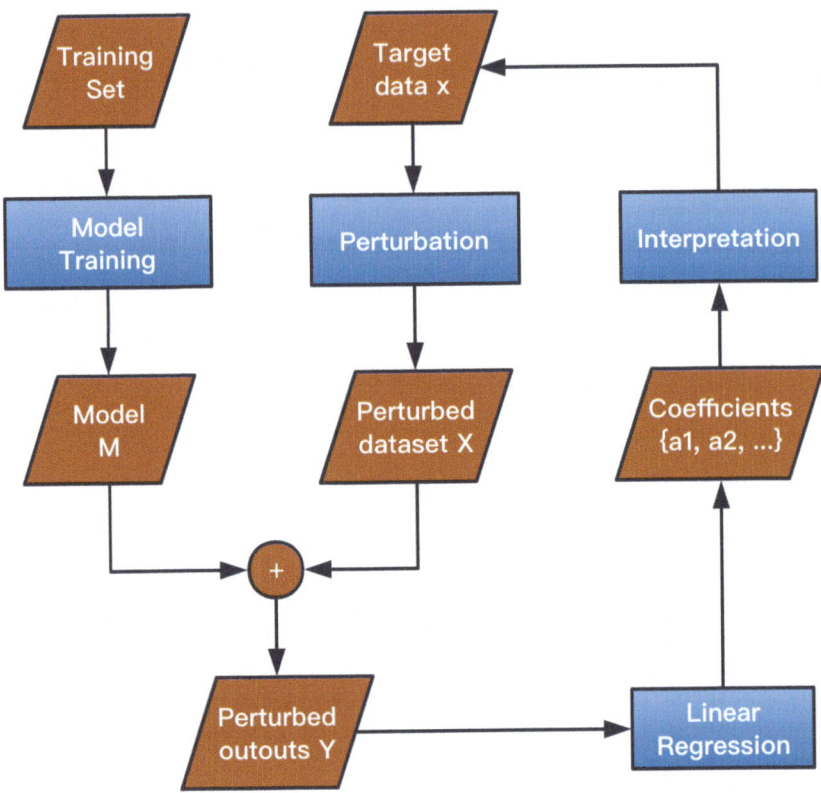

Fig. 3 The proposed malware detection framework, which consists of four major activities: model training, perturbation, linear regression, and outcome interpretation

depth. It represents the recorded values of w traced signals over d clock cycles. We split each column as a single feature component, i.e., values of all traced signals within one single cycle. Next, we apply explainable machine learning for malware detection. Figure 3 shows the overview of proposed method that consists of four major tasks. The first task is *model training*, where we normally train a machine learning classifier M using collected traces. The second task performs *perturbation*. For an input instance $\mathbf{x}$ we want to explain, perturb $\mathbf{x}$ randomly to generate artificial input dataset $\mathbf{X} = \{\mathbf{x}_1, \mathbf{x}_2, \ldots\}$, and feed them to M to obtain corresponding model output $\mathbf{Y} = \{M(\mathbf{x}_1), M(\mathbf{x}_2), \ldots\}$. Also, in order to guarantee the accuracy of the subsequent regression algorithm, we need to eliminate isolation points in $\mathbf{Y}$. The third task applies *linear regression* on $\mathbf{X}$ and $\mathbf{Y}$ to obtain a linear regression function. The goal of the last task is to perform *outcome interpretation*. Specifically, the top features ranked by the magnitude of coefficients will provide users the crucial timing information of malware. The remainder of this section describes these four tasks in detail.

3.1 Model Training

Hardware-assisted malware detection techniques should monitor the behavior of software at runtime. Therefore, relying on single-cycle data is not effective since malicious behavior usually consumes several sequential cycles. Moreover, single-cycle-based strategies are likely to mispredict a benign software as malicious. This is due to the fact that malware also contains normal operations, and considering these benign operations as important features of malware can lead to misclassification. A well-designed preprocessing strategy can mitigate this mistake by filtering overlapped common behaviors shared by both, but the difficulty of designing such a strategy is extremely high and there is no guarantee that it can be performed in all cases. Therefore, an ideal machine learning model for our task should satisfy the following two properties:

(1) Ability to accept time series type data as input.
(2) Ability to make decisions utilizing potential information concealed in consecutive adjacent inputs.

We utilize *Recurrent Neural Network (RNN)* training to tackle this problem. Algorithm 1 outlines the training procedure. In order to explain the working principles of the algorithm, we need to describe RNN as well as its importance in the context of malware detection. RNN is powerful in handling sequential input data. A classic structure of RNN is shown in Fig. 4.

In the picture, **A** represents the neural network architecture, where $x_0, x_1, x_2, \ldots$ means the time series inputs and h_is are the outputs of hidden layers. As we can see from the left side of the figure, instead of finishing the input–output mapping in one forward pass, the RNN accepts sequential inputs. For each single input x_i, RNN not only provides immediate response h_i but also stores the information of the current input by updating the architecture itself. On the right side of the figure, information corresponding to the previous step will also be fed into the architecture to supply extra information by unrolling the RNN structure. For trace data-based malware

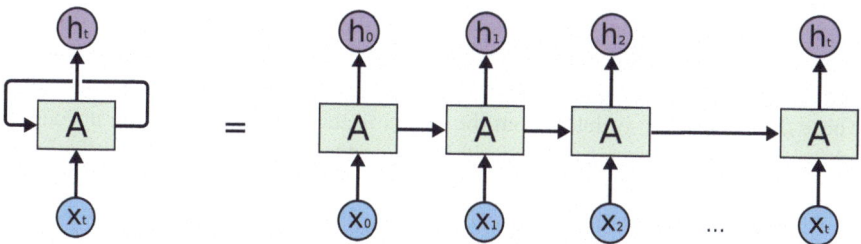

Fig. 4 A simple diagram of Recurrent Neural Network (RNN). The RNN structure allows the output to affect the inputs in subsequent iterations. Their ability to use internal state (memory) to process arbitrary sequence of inputs makes them applicable to handle sequential data

detection, we can directly set each column of trace table as inputs and accept the hidden state of final stage, i.e., h_t, as the final output.

RNN is suitable for handling time-sequential data, but it is not guaranteed to learn features from adjacent inputs. For time series inputs, considering inputs in groups and training the model to make decision based on the co-occurrence of sequential features are crucial. This can be achieved by appending a penalty term to the loss function, and this term can force models to group adjacent elements together from input feature map. The loss function with penalty for RNN model (Fig. 4) can be written as

$$J = \frac{1}{N} \sum_{i=1}^{N} L(\mathbf{A}(\mathbf{x}_i), \mathbf{y}_i) + \frac{\lambda}{2} \sum_{k=1}^{t} ||h_i - h_{i-1}||$$

where $\mathbf{A}$ is the model, $\mathbf{x}_i$ is a training sample, the label of $\mathbf{x}_i$ is denoted as $\mathbf{y}_i$, the total number of training samples in a batch is denoted as N, and L is the dissimilarity measurement which is frequently selected to be cross entropy for classifiers. Aside from these regular terms, we introduce a penalty term $\sum_{k=1}^{t} ||h_i - h_{i-1}||$, which tries to minimize the difference between the hidden state outputs of each time step. This is to restrict the impact brought by one single clock cycle input and prevent the machine learning model from updating its inner feature map too significantly unless it produces a relatively long sequence of similar patterns. Based on the assumption that malicious behavior happens in multiple sequential cycles instead of just one, this training scheme enables RNN to take adjacent inputs as groups for gathering information and making decisions.

3.2   Perturbation and Outlier Elimination

Once we have the well-trained model, we can start to perturb the target input $\mathbf{x}$ to generate corresponding perturbed output dataset $\mathbf{Y}$. This is achieved by randomly flipping several bits in target input $\mathbf{x}$. However, the raw output $\mathbf{Y}$ cannot be directly applied to regression algorithm in the next step. This is due to the fact that random perturbation can generate anomalous data, such as data points with extreme value. These data points are isolated from the others in the cluster, so they can introduce a huge deviation in regression algorithms. To address this, we need to efficiently remove isolated points in $\mathbf{Y}$.

We deployed a random voting algorithm to achieve this goal. The basic idea is to cut a data space with a random hyperplanes, and two subspaces can be generated at a time. We continue to randomly select hyperplane to cut subspaces obtained in the previous step, and the process continues until each subspace contains only one data point. Intuitively, we can find that those clusters with high density will not be

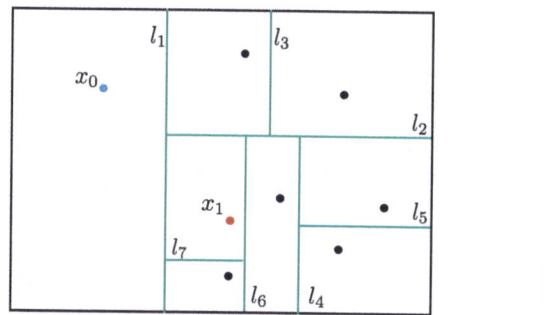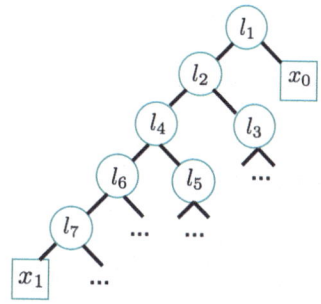

Fig. 5 Finding outliers using isolation forest. In an isolation forest, randomly sub-sampled data is processed in a tree structure based on randomly selected features. The samples that travel deeper into the tree are less likely to be anomalies as they require more cuts to isolate them. Similarly, the samples that end up in shorter branches indicate anomalies as it is easier for the tree to separate them from other observations

entirely dismembered until they are cut many times, but those in the low density regions are separated out much earlier.

Figure 5 shows a simple example. If we want to isolate x_0, we need to draw l_1, i.e., cut the space one time, while x_1 needs a lot more. So x_0 is more likely to be an outlier than x_1. Note that the process of cutting space can be naively represented by a binary tree as shown in Fig. 5.

In general, a threshold θ is applied to categorize isolated and clustered points. For each data point, we check the depth of it inside the binary tree. A point is considered as isolated once the depth exceeds the given threshold. These isolated points are more likely to be extreme value data points and should not be used by the regression algorithm. Eliminating them is likely to improve the accuracy of regression. In order to ensure reliability, we repeat this procedure for several times to obtain a forest of trees and let them vote for the final decision.

3.3 Linear Regression

A linear regression algorithm allows us to approximate locally nonlinear relationship with proper precision. Formally, given a dataset $\{\mathbf{y}, \mathbf{x}_1, \mathbf{x}_2, \ldots, \mathbf{x}_n\}$, where n is the number of samples, linear regression takes the following form by appending error variable ϵ:

$$\mathbf{y} = \sum_{i=1}^{n} a_i \mathbf{x}_i + \epsilon$$

where a_is are model parameters, and the goal is to minimize ϵ as much as possible. The simplest scenario occurs when $\mathbf{y}$ and every $\mathbf{x}_i$ are real numbers. In our case, the input is the $w * d$ trace table $\mathbf{X}$ as mentioned before. Since we treat each column of this table as an individual input feature, we have $\mathbf{X} = [\mathbf{x}_1\ \mathbf{x}_2\ \dots\ \mathbf{x}_d]$, where each $\mathbf{x}_i$ is a vector in the size of $w * 1$. We choose $\mathbf{y}$ as the output of last hidden state, i.e., h_t in Fig. 4 which is also a $w * 1$ vector. This leads to an optimization problem:

$$\arg \min_{\mathbf{a}} ||\mathbf{Xa} - \mathbf{y}||_2$$

where $\mathbf{a} \in \mathbb{R}^d$ is $[a_1\ a_2\ \dots\ a_d]^T$, i.e., coefficients to be solved. This is a common convex optimization problem and its solution can be obtained by *least squares* which gives

$$\mathbf{a} = (\mathbf{X}^t\mathbf{X})^{-1}\mathbf{X}^t\mathbf{y}$$

Unfortunately, this method cannot be directly applied to solve our task. First, this theoretical solution exists only when $\mathbf{X}^t\mathbf{X}$ is invertible (full rank), which is not satisfied for most of the time. Second, even when $\mathbf{X}^t\mathbf{X}$ is full rank, linear regression assumes input vectors are *independent*, otherwise it will produce unreliable results when any two of $\mathbf{x}_i$ (columns) are highly correlated. Specifically, assume that the regression function is computed to be $\hat{\mathbf{y}} = a\mathbf{x}_1 + b\mathbf{x}_2 + c\mathbf{x}_3 + d$, where $\mathbf{x}_1$ and $\mathbf{x}_2$ are highly related features and they are very close to each other. Then there is a canceling effect between a and b. Increasing a by certain amount while decreasing b by the same amount at the same time will not lead to drastic change in $\hat{\mathbf{y}}$. This can cause high variance of computation results for coefficients. The problem becomes ill-posed since the absolute value of a and b can vary significantly under different computing procedure or initial conditions. Then the comparison between $|a|$, $|b|$ and $|c|$ is not useful, and therefore, the interpretability of the model is greatly reduced. Since adjacent columns in trace table are sequential records of signal values within a short duration, violation of this independence assumption is likely to happen. Specifically, we applied *ridge regression*, which is an improved least squares estimation method, and the fitness of correlated data is stronger than general regression. Ridge regression is achieved by appending one extra penalty term to the optimization problem:

$$\arg \min_{\mathbf{a}} ||\mathbf{Xa} - \mathbf{y}||_2 + \lambda||\mathbf{a}||_2$$

Intuitively, a size constraint is imposed to restrict the absolute value of all coefficients, which alleviate the problem of high variance of coefficients. Moreover, notice

$$\arg \min_{\mathbf{a}} ||\mathbf{Xa} - \mathbf{y}||_2 + \lambda||\mathbf{a}||_2 \rightarrow \arg \min_{\mathbf{a}} ||(\mathbf{X} - \lambda\mathbf{I})\mathbf{a} - \mathbf{y}||_2$$

Replacing **X** with **X** − λ**I** is a general way to avoid the problem for **X** being singular matrix. Also, data was centralized and the problem of high variance is alleviated. Therefore, with ridge regression, coefficients obtained are more reliable and fit better for our dataset, which has high correlation.

3.4 Outcome Interpretation

Once coefficients of regression are obtained, we can derive the importance ranking and then interpret it into meaningful information in the context of malware detection. The top features come with large coefficients that are likely to be related to the malicious behavior. Next, we can check the clock cycle distribution of these top features. It is expected to provide us with extra information about the malware. For example, if we observe an adjacent cluster of top features, then the time slot within which they reside shall provide a general indication of time information about when malicious behavior happened. Similarly, if clock cycle numbers are periodically separated, the detected malware is likely to repeat its malicious activity periodically. Typical malware acting like this usually works in a client–server mode, where the client program steals private data and sends message to the hacker's server in a periodic interval. For a closer look, we can split the table into rows and go through the same process. This will lead to the identification of trace signal values that are most likely relevant to the malicious behavior, which in turn will lead to malware localization as demonstrated in the next section.

4 Experiments

This section is organized as follows. First, we outline the experimental setup and describe various malware and benign benchmarks. Next, we discuss various data acquisition techniques as well as applied ML models. Finally, we present results in terms of malware detection accuracy, time efficiency, and outcome interpretation.

4.1 Experimental Platform

We ran malicious and benign programs on the Xilinx Zynq-7000 SoC ZC702 evaluation board as shown in Fig. 6. This board integrates double ARM Cortex-A9 cores. We installed $xilinx - zc702 - 2017\_3 : 4.9.0 - xilinx - v2017.3$, a Linux kernel image for the ZC702 evaluation board generated using PetaLinux, to the board using the provided 8 GB SD Card. To view the contents of internal signal values, we link the board to the System Debugger in Xilinx SDK version 2017.3, which uses a hardware server to allow us to compile and run these programs on the

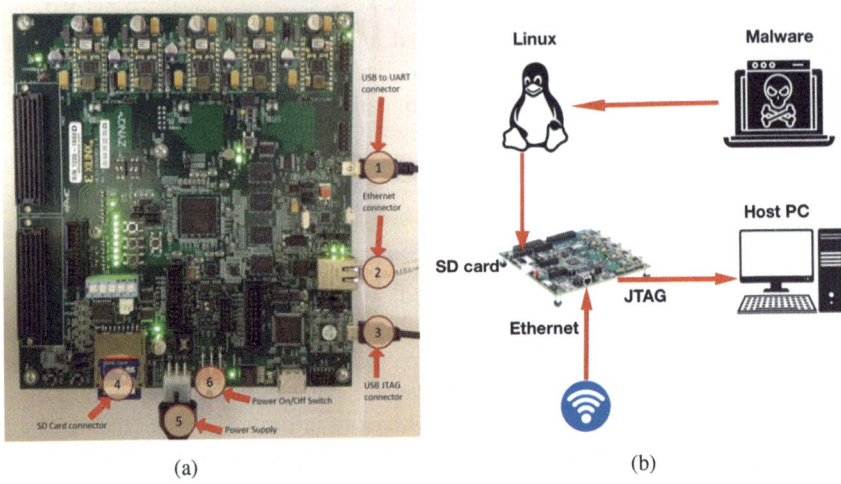

(a)   (b)

Fig. 6 Experimental platform. (**a**) ZYNQ SoC board. (**b**) Platform layout

board while monitoring traced signal values. This configuration involves connecting the board to a host computer running Xilinx SDK using Ethernet (to run programs using the System Debugger) and JTAG (to dump register data provided by the SDK and set up the Ethernet connection).

4.2 Malware and Benign Benchmarks

In this study, we consider a wide variety of malware families [2] including the following three popular ones: *BASHLITE Botnet*, *PNScan Trojan*, and *Mirai Botnet*.

- BASHLITE, also known as Gafgyt or LizardStresster, is a malware family targeting Linux systems. BASHLITE infiltrate in IoT devices and these poisoned devices will be used to manipulate large-scale distributed denial-of-service (DDoS) attacks. It uses Shellshock to gain a foothold on vulnerable devices and then remotely executes commands to launch DDOS attacks and download other files to the compromised device. It works in a client–server mode where poisoned devices keep sending requests to a remote server checking for possible update releases or malicious requests.
- PNScan is an open-source Linux Trojan which can infect devices with ARM, MIPS, and PowerPC architectures. This Trojan or applications with this Trojan embedded can invade network devices. This malicious program has only one goal—obtain the router's access password through brute force. If the intrusion is successful, the Trojan will load a malicious script into the router which will

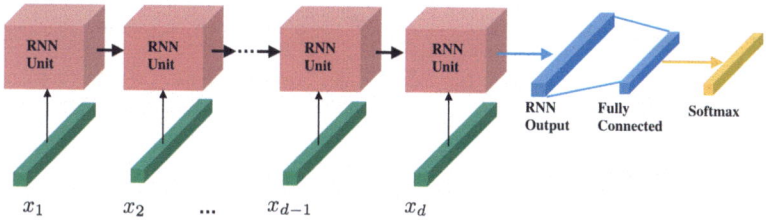

Fig. 7 The implemented RNN classifier architecture

download the corresponding backdoor program according to the router's system architecture.

- Mirai is an upgraded variant of BASHLITE. Mirai can efficiently scan IoT devices and infect fragile devices like the ones encrypted with default factory settings or weak passwords. After being infected by this malware, the device becomes a botnet robot and launches a high-intensity botnet attack under the command of a hacker.

The benign program benchmarks include system binaries such as *ls*, *mkdir*, *ping*, *netstat*[1]. The traced values gathered by running both malware and benign programs on the hardware board are utilized as inputs to our classifier, as described next (Fig. 7).

4.3 Data Acquisition

As shown in Fig. 8, we have collected data through the following three avenues.

Hardware Performance Counters (HPC) To gather HPC data, we make a serial connection to the target device and run both the malicious and benign programs with the *perf* record command. Within the perf command, we set various HPC values (such as branch mispredictions, number of branches, number of loads, etc.) as traced events.

Embedded Trace Buffer (ETB) To obtain ETB traces, we run malicious programs on the target system while connected to Xilinx SDK through Ethernet. Within Xilinx SDK, we add breakpoints to various code locations and dump trace data at these breakpoints. Our studies have shown that tracing register values provide better insight compared to tracing other signals.

Synergistic Combination of HPC and ETB As shown in Fig. 8, we dump ETB data only when our ML models find any suspicious HPC values. To associate regions of HPC values with ETB values, we add dynamic tracepoints (using the perf

---

[1] Ping and netstat are important since our malware are botnets.

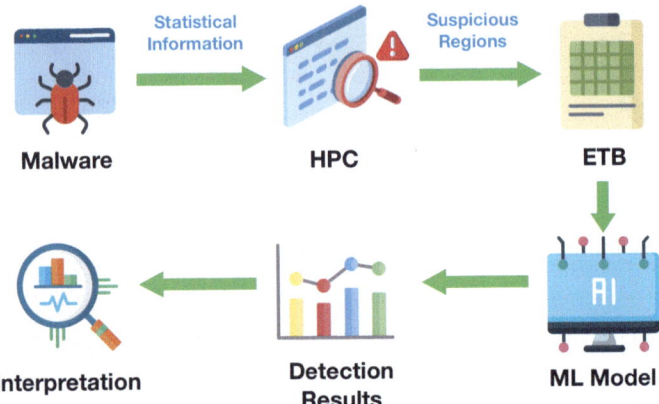

Fig. 8 Malware detection using both embedded trace buffer (ETB) data and hardware performance counter (HPC) values

probe functionality) to various functions in the malicious programs. The dynamic tracepoints traced during HPC data generation allow us to associate regions of HPC data with ETB data. By remembering which function was called near a suspicious HPC data region, we can place a breakpoint in the ETB region associated with suspicious HPC activity. In other words, we can generate ETB data corresponding to specific HPC values.

4.4 RNN Classifier

Figure 7 presents the architecture of implemented classifier. The proposed classifier utilizes the structure outlined in Fig. 4. Here $\{x_1, x_2, \ldots x_d\}$ are values of traced signals in d different clock cycles. After passing through RNN units, the outputs are fed into a fully connected layer to achieve dimension reduction. Finally, the softmax layer takes the reduced outputs from fully connected layer to produce classification labels.

As for RNN units, there are three most widely applied prototypes, *traditional* (also known as *vanilla units*), *GRU* [7], and *LSTM* [25]. We applied GRU and the reason for making this choice is threefold.

1. Traditional units suffer from the *vanishing gradient* [12] and *exploding gradient* [22] problem. It usually contains an extremely long chain of RNN units for handling complicated task. When performing *backpropagation* [24], if the initial gradient is less than 1, then the gradient at the last moment will disappear and vice versa. Both situations will lead to failure in the training process.
2. LSTM adopted a *gate* mechanism [7] to solve the vanishing gradient problem. Meanwhile, the gate mechanism provides feature filtering, saving useful features

and discarding useless features, which greatly enriches the information representation capacity of the model. However, LSTM suffers from high computation time cost, which is the key barrier for us to adopt it in our study since we want a low-latency malware detection technique.

3. GRU also utilizes a gate mechanism to solve the gradient-related problems and can be regarded as a simplified version of LSTM. By merging overlapped gates and hidden states, the model structure is much simpler than LSTM. Simpler structure brings fewer parameters and faster convergence. GRU takes much less time, which can greatly speed up the training process.

4.5 Evaluation: Accuracy

We first compare the accuracy of our proposed approach with the state-of-the-art hardware-assisted malware detector, PREEMPT [4]. PREEMPT utilizes embedded trace buffer to help reducing latency and overcoming malware equipped with obfuscation. PREEMPT utilizes two types of implementation: random forest (PREEMPT_RF) and decision tree (PREEMPT_DT) .

We run both malicious and benign software on a given hardware platform. We executed a total of 367 programs (including both malicious and benign ones) and all the traced data were mixed up and further split into training (80%) and test (20%) sets after labeling. The total training epochs are 200 for every model and we plot test accuracy every 10 epochs. The performance of all methods is depicted in Fig. 9.

Figure 9 compares the prediction accuracy of our approach with PREEMPT_RF and PREEMPT_DT. As we can see, our proposed method provides the best malware detection accuracy. The PREEMPT appeared fragile in the face of PNScan, with an average of 62.7% accuracy for DT and 76.9% for RF, while the proposed method provided an average accuracy as high as 91.4%. For BASHLITE, both proposed method and RF performed well and the best accuracy of our method is 98.9%. For Mirai, our proposed method achieved 97.5% accuracy, while PREEMPT attained a maximum accuracy of 92.5% with RF. Note the inferior performance of PREEMPT_DT in Mirai dataset.

If we omit malware and test models on traced data gathered from benign software only, Fig. 9d shows the false positive rate (FPR) of all three methods. The diagram illustrates the major drawback of PREEMPT, and it possesses an average FPR as high as 25.9% with RF and 31.6% with DT. In other words, it is very likely to mispredict a benign software as malware. Tested benign software samples also execute Linux system binaries like *netstat* and *ping*, which are also frequently executed by botnet malware. Since the PREEMPT cannot analyze time-sequential data, it failed to recognize a benign execution of these binaries with the help of context and produced wrong predictions. In contrast, our framework obtained FPR as low as 3.4%. Next we discuss how our framework achieved promising performance with the help of explainable machine learning.

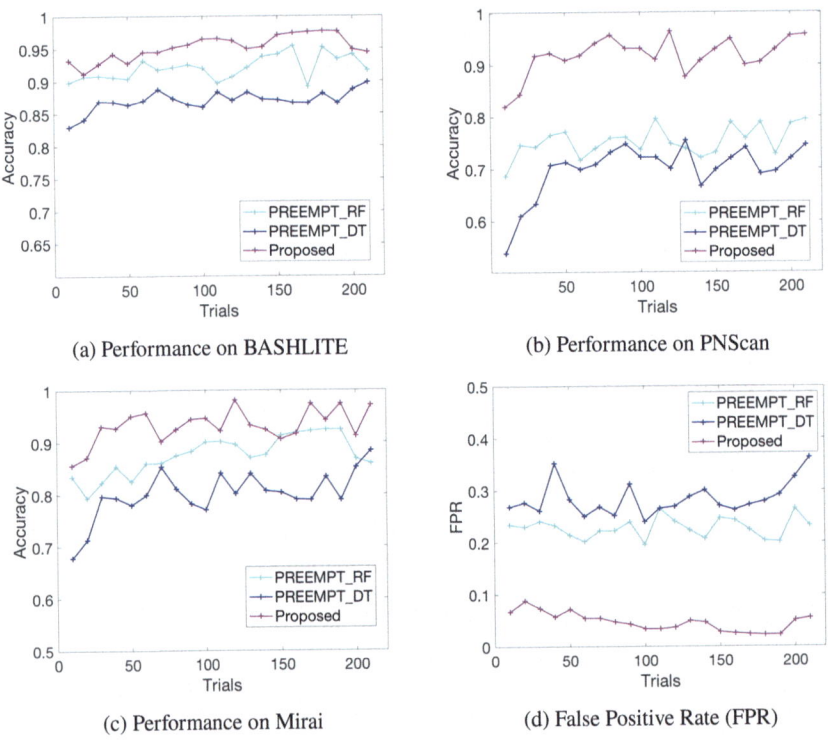

(a) Performance on BASHLITE   (b) Performance on PNScan

(c) Performance on Mirai   (d) False Positive Rate (FPR)

Fig. 9 Performance of machine learning models: (**a**)–(**c**) for various malware and (**d**) for benign benchmarks

4.6 Evaluation: Outcome Interpretation

We also evaluated the performance of our method by interpreting the contribution factor for the classification results. For clock cycle-related analysis, an example of executing BASHLITE's client on host machine is shown in Fig. 10.

Figure 10 shows a snapshot of the trace table, where each row represents the values in a register in specific clock cycles (each column represents a specific clock cycle). The proposed method computed the corresponding contribution factor of each clock cycle toward the RNN output using linear regression, which is shown as weights in the last (colored) row. As we can see, the weight of C_4 is significantly larger than the others. This immediately indicates the clock cycle of malicious behavior. By tracing the execution, we find out C_4 points to the timestamp before the start of function "*processCmd*" in BASHLITE, which is the most important function of BASHLITE to perform its malicious functionality. In other words, this is the starting point and exact reason for recognizing this program as malware.

Another example of outcome interpretation is shown in Fig. 11, where we measure the contribution of each traced register signal. The given data is the

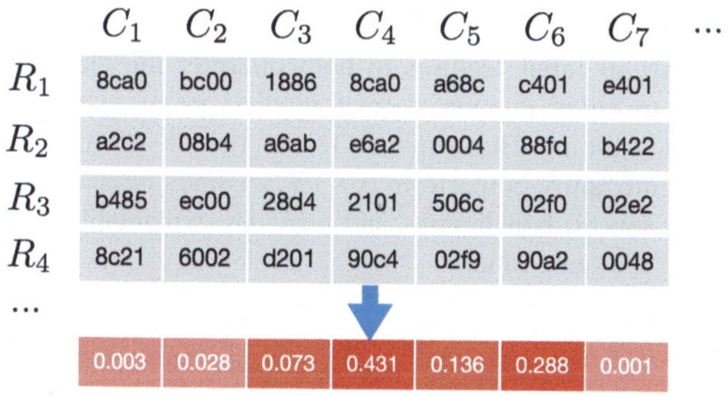

	C_1	C_2	C_3	C_4	C_5	C_6	C_7	$\cdots$
R_1	8ca0	bc00	1886	8ca0	a68c	c401	e401	
R_2	a2c2	08b4	a6ab	e6a2	0004	88fd	b422	
R_3	b485	ec00	28d4	2101	506c	02f0	02e2	
R_4	8c21	6002	d201	90c4	02f9	90a2	0048	
$\cdots$								

0.003	0.028	0.073	0.431	0.136	0.288	0.001

Fig. 10 Interpretation of BASHLITE client's traced signals

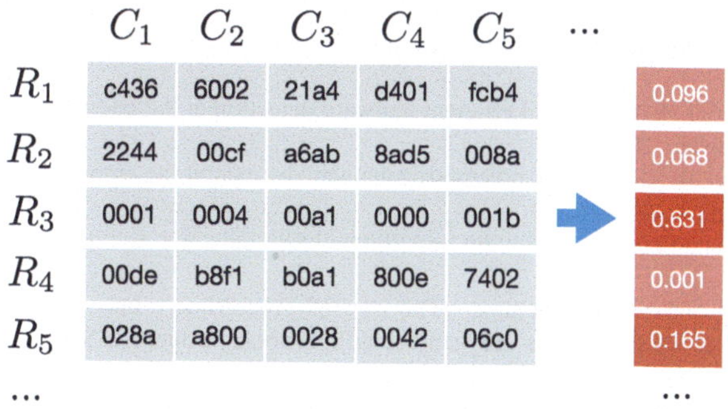

	C_1	C_2	C_3	C_4	C_5	$\cdots$	
R_1	c436	6002	21a4	d401	fcb4		0.096
R_2	2244	00cf	a6ab	8ad5	008a		0.068
R_3	0001	0004	00a1	0000	001b		0.631
R_4	00de	b8f1	b0a1	800e	7402		0.001
R_5	028a	a800	0028	0042	06c0		0.165
$\cdots$							$\cdots$

Fig. 11 Interpretation of Mirai bot's traced signals

trace table of executing Mirai's bot on host machine. This time we evaluate the contribution row by row, and the result is listed on the right side of the trace table.

As we can see, register R_3 is recognized as the most important factor. Here R_3 is storing the variable "*ATTACK_VECTOR*" in Mirai. This variable records the identity of attack modes, based on which the bot takes relative actions to perform either a UDP attack or a DNS attack. This attack-mode flag is the most important feature of a majority of malware bot programs, and the proposed method successfully extracted it from the traces to illustrate the reason for making this prediction.

5 Summary

Malicious software (malware) is a serious threat to modern computing systems. Existing software-based solutions are not effective in the face of attacks with obfuscation or other evasive techniques. Recent hardware-based detection techniques are promising, but their detection accuracy can still be improved. Moreover, the classification results cannot be interpreted in a meaningful way. In this chapter, we tried to address these limitations by developing a regression-based explainable machine learning algorithm. The proposed approach is able to find the major contributors among all input features to help interpret the classification results, which is utilized to either support correct predictions or justify mispredictions through adversarial training. Experimental results demonstrated that the proposed framework can significantly outperform (97.7% accuracy on average) state-of-the-art approaches on several benchmarks and provides explainable interpretation on detection results at the same time.

References

1. *Malware Obfuscation Techniques: A Brief Survey*. IEEE Computer Society, 2010.
2. Kishore Angrishi. Turning Internet of Things(IoT) into Internet of Vulnerabilities (IoV). *CoRR*, 2017.
3. Daniel Arp et al. Effective and efficient malware detection at the end host. In *18th USENIX*, Montreal, Quebec, 2009.
4. Kanad Basu et al. PREEMPT: preempting malware by examining embedded processor traces. In *DAC*, page 166, 2019.
5. Zahra Bazrafshan et al. A survey on heuristic malware detection techniques. In *ICIKF*, pages 113–120, 2013.
6. Kelly Bissell and Larry Ponemon. The cost of cybercrime. https://www.accenture.com/t20190305T185301Z__w__/us-en/_acnmedia/PDF-96/Accenture-2019-Cost-of-Cybercrime-Study-Final.pdf#zoom=50, 2019.
7. Junyoung Chung et al. Empirical evaluation of gated recurrent neural networks on sequence modeling. *CoRR*, 2014.
8. George E. Dahl, Jack W. Stokes, Li Deng, and Dong Yu. Large-scale malware classification using random projections and neural networks. In *ICASSP*, pages 3422–3426, 2013.
9. John Demme et al. On the feasibility of online malware detection with performance counters. In *The 40th Annual ISCA*, pages 559–570, 2013.
10. Timon Gehr et al. AI2: safety and robustness certification of neural networks with abstract interpretation. 2018.
11. Kathrin Grosse et al. Adversarial perturbations against deep neural networks for malware classification. *CoRR*, 2016.
12. Sepp Hochreiter. The vanishing gradient problem during learning recurrent neural nets and problem solutions. *International Journal of Uncertainty, Fuzziness and Knowledge-Based Systems*, 6(2):107–116, 1998.
13. Nwokedi Idika and Aditya Mathur. A survey of malware detection techniques. *Purdue University*, 03 2007.
14. Grégoire Jacob, Hervé Debar, and Eric Filiol. Behavioral detection of malware: from a survey towards an established taxonomy. *Journal in Computer Virology*, 4(3):251–266, 2008.
15. Suman Jana and Vitaly Shmatikov. Abusing file processing in malware detectors for fun and profit. In *IEEE S&P*, 2012.
16. Nick L. Petroni Jr. et al. Copilot - a coprocessor-based kernel runtime integrity monitor. In *Proceedings of the 13th USENIX Security Symposium*, pages 179–194, 2004.
17. Mikhail Kazdagli et al. Quantifying and improving the efficiency of hardware-based mobile malware detectors. In *49th Annual IEEE/ACM MICRO*, pages 37:1–37:13, 2016.
18. Yann LeCun, Yoshua Bengio, and Geoffrey E. Hinton. Deep learning. *Nature*, 521(7553):436–444, 2015.
19. A. Moser, C. Kruegel, and E. Kirda. Limits of static analysis for malware detection. pages 421–430, Dec 2007.

20. Zhixin Pan and Prabhat Mishra. Automated detection of spectre and meltdown attacks using explainable machine learning. In *2021 IEEE International Symposium on Hardware Oriented Security and Trust (HOST)*, pages 24–34. IEEE, 2021.
21. Zhixin Pan, Jennifer Sheldon, and Prabhat Mishra. Hardware-assisted malware detection and localization using explainable machine learning. *IEEE Transactions on Computers*, 71(12):3308–3321, 2022.
22. Razvan Pascanu, Tomas Mikolov, and Yoshua Bengio. Understanding the exploding gradient problem. *CoRR*, 2012.
23. Marco Túlio Ribeiro, Sameer Singh, and Carlos Guestrin. "why should I trust you?": Explaining the predictions of any classifier. In *Proceedings of the 22nd ACM SIGKDD*, 2016.
24. David E. Rumelhart, Geoffrey E. Hinton, and Ronald J. Williams. Learning representations by back-propagating errors. *Nature*, 323:533–536, 1986.
25. Hasim Sak et al. Long short-term memory based recurrent neural network architectures for large vocabulary speech recognition. *CoRR*, abs/1402.1128, 2014.
26. Joshua Saxe and Konstantin Berlin. Deep neural network based malware detection using two dimensional binary program features. In *10th MALCON*, pages 11–20, 2015.
27. Karen Simonyan, Andrea Vedaldi, and Andrew Zisserman. Deep inside convolutional networks: Visualising image classification models and saliency maps. 2014.
28. Qinglong Wang et al. Adversary resistant deep neural networks with an application to malware detection. In *Proceedings of the 23rd ACM SIGKDD*, pages 1145–1153, 2017.
29. Xueyang Wang et al. Hardware performance counter-based malware identification and detection with adaptive compressive sensing. *ACM TACO*, 13(1):3, 2016.
30. Xueyang Wang and Ramesh Karri. NumChecker: detecting kernel control-flow modifying rootkits by using hardware performance counters. In *DAC*, pages 79:1–79:7, 2013.
31. Hasini Witharana and Prabhat Mishra. Speculative load forwarding attack on modern processors. In *IEEE/ACM International Conference on Computer-Aided Design*, pages 1–9, 2022.

Spectre and Meltdown Detection Using Explainable AI

1 Introduction

In this chapter, we discuss two software attacks that exploit the speculative execution properties in modern processors, popularly known as Spectre and Meltdown attacks. Modern processors utilize advanced architectural features, such as branch prediction and out-of-order execution, to significantly improve the performance of computing devices. As shown in Fig. 1, processors can perform parallel processing of predicted tasks with excess system resources by utilizing speculative execution. However, these performance-enhancing techniques introduce security vulnerabilities that are exploited by Spectre [16] and Meltdown [21] attacks. Spectre attack abuses the "branch prediction" capability to break devices' memory isolation capabilities, while Meltdown attack exploits the vulnerability arising from "out-of-order execution." These attacks enable a malicious process to gain unauthorized access to memory locations. For example, the Meltdown attack can dump kernel memory at a speed of 503 KB/s [21].

While there are a large number of existing efforts [12, 14, 35, 37] to defend against these attacks, existing approaches focus on mitigation techniques that can lead to unacceptable performance degradation. For example, some techniques recommend either occasional shut down of speculative execution or clearing of branch target buffer during context switching. There are recent efforts that utilize machine learning (ML) for the detection of Spectre and Meltdown attacks [2]. However, ML-based approaches have two inherent weaknesses. They cannot detect evasive Spectre and Meltdown attacks that rely on obfuscation techniques or other deviation capabilities [19]. Moreover, due to the black-box nature of ML models, the prediction results cannot be interpreted for debugging or improving the model accuracy [27].

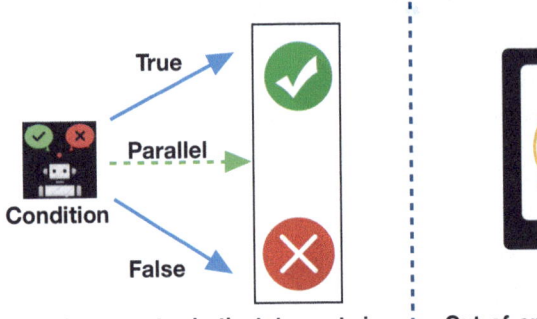

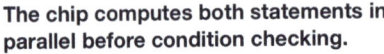

The chip computes both statements in parallel before condition checking.

Out-of-order execution to avoid waiting caused by fetching the next instruction.

Fig. 1 Modern processors improve performance using speculative execution methods, such as branch prediction (left) and out-of-order execution (right)

1.1   Threat Model

In this chapter, we make the following assumptions about the attacker's objectives, knowledge, and attack modes, which are consistent with related efforts in Sect. 2.

Objectives The attacker's objective is to cause information leakage by revealing values in concealed memory locations without authorization.

Knowledge We assume that the attacker is aware of specific signatures (patterns) used by existing detection techniques. We also assume that the adversary has sophisticated knowledge of security vulnerabilities as well as information about the target architecture and operating system.

Attack Modes The adversary will use well-crafted Spectre and Meltdown attack codes to attack the target device. The malicious process should be able to raise an exception during program execution and perform cache timing attack. We also assume that the adversary can exploit the knowledge of specific patterns used by detection techniques to devise evasive Spectre and evasive Meltdown.

1.2   Motivation

Existing efforts [2, 7, 13, 19, 20, 33] are promising to prevent Spectre and Meltdown attacks but suffer from two inherent limitations: high overhead and poor robustness. The high overhead issue arises from the fact that passive prevention of Spectre attacks involves selectively turning off speculative execution [7, 20] to prevent possible attacks, which results in a significant reduction in performance. Moreover, software patches that mitigate Meltdown [22] can introduce unacceptable overhead [26]. Architectural alterations [13, 33] also increase the burden on the pipeline, and detection techniques take considerable time to complete before normal execution can proceed. Consequently, these approaches result in significant overhead in terms of both performance and additional design complexity.

The poor robustness problem arises from the reliance on hardware-based detection methods, which are preferred over software-based solutions because of their low latency. Most of these approaches utilize *Hardware Performance Counter* (HPC) values that monitor hardware events such as cache misses and branch misprediction in microprocessors. HPCs are beneficial since Spectre and Meltdown attacks leave noticeable traces in HPCs during triggering of exception and cache access measurements and has been widely applied in malicious attack detection [29, 30]. As a result, machine learning can detect malicious attacks by observing specific patterns in HPC values. However, ML-based detection is vulnerable to evasive methods [19], where benign functions are invoked between malignant payloads, and dummy instructions are inserted to increase specific HPC values to fool the detectors. Existing detection methods fail miserably (less than 60% detection rate), which is comparable to a random guess.

In this chapter, we discuss a hardware-assisted detection framework for Spectre and Meltdown attacks, leveraging the power of explainable machine learning. We first provide theoretical analysis that establishes a clear correlation between hardware events and the features inherent to Spectre and Meltdown attacks. This correlation enables us to achieve high robustness against obfuscation. Next, we discuss the use of hardware events as time-sequential inputs to mitigate the effect of misprediction caused by obfuscation techniques. This makes our trained ML model resistant to evasive attacks. Then, we explore the effectiveness of long short-term memory (LSTM) as well as ensemble boosting to improve the training efficiency of ML models. Finally, we investigate the usefulness of diverse explainable ML models, including model distillation and Shapley values, for interpreting the prediction results.

2 Background and Related Work

There are many research efforts in efficient detection and mitigation of security vulnerabilities [6, 8, 10, 23, 24, 28, 29, 36]. This section provides background on Spectre and Meltdown attacks. It also highlights the need for explainable machine learning in detecting Spectre and Meltdown attacks.

2.1 Spectre and Meltdown Attacks

The security of an operating system is fundamental, as it must prevent user programs from accessing the kernel or any other programs' memory locations. If a user program attempts an illegal access, the CPU should detect the permission violation

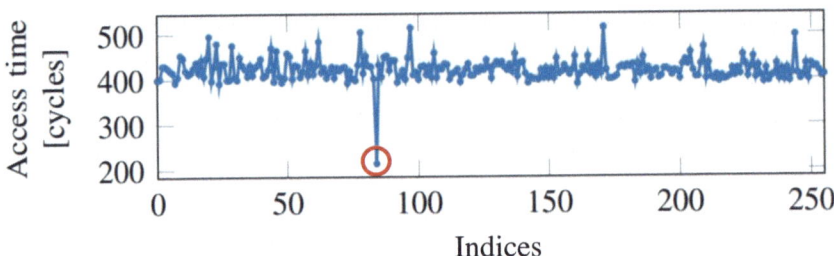

Fig. 2 During cache timing attack, an adversary can identify the data corresponding to the index with the shortest access time since it is likely to be pre-stored in the cache [21]

and terminate the program. However, transient execution attacks [4], also known as speculative execution attacks, exploit the behavior of modern processors to case information leakage during this process. Spectre and Meltdown are two of the most famous transient execution attacks, where the attacker can take advantage of speculative execution such as cache state changes and timing variations, to infer the values of sensitive data.

Listing 1 illustrates a simple template for a Meltdown attack:

```
mov rax byte[x]     // illegal access
shl rax 0xC         // page alignment
mov rcx rbx[rax]    // probe data
```

Listing 1 Example Meltdown attack

Here, "rax," "rbx," and "rcx" are register names, and "byte[x]" is a private memory location that will raise an exception upon illegal access. The left shift by 12 bits in the second instruction multiplies "rax" by the page size (4096). Ideally, "rax" should be cleared before executing subsequent instructions, but speculative execution causes the second and third instructions to be partially executed before the exception handling takes effect. Additionally, modern cache designs bring "rax" into the cache to hide the latency of subsequent accesses if it is not already there. While "rax" is cleared by exception handling, the cache is not immediately flushed, temporarily storing the latest illegal access information. As shown in Fig. 2, attackers can use a cache-based side-channel attack to restore this address by traversing the array headed by "rbx" and measuring the access time, with the page having the shortest access time addressed by "rax."

Similarly, a Spectre attack exploits branch prediction, as shown in Listing 2:

```
if(x < arr1_size);      // boundary check
y = arr2[arr1[x]*4096]; // array access
```

Listing 2 Example Spectre attack

If "x" is out of range, the second line should not execute. However, due to branch prediction, it will still be pre-executed, leaving traces in the cache that can be retrieved through cache timing attack. Compared to Meltdown, Spectre has a wider attack range, making it more dangerous.

2.2 *Explainable Machine Learning*

Explainable ML aims to provide interpretable explanations for the results of ML models. During the testing phase, given an input instance **x** and a traditional ML model, the classifier will generate a corresponding output **y** for **x**. Explanation techniques then attempt to illustrate why instance **x** is classified into **y**. This involves identifying a set of crucial features inside **x** that significantly contribute to the classification result. If the selected features are interpretable by human analysts, they can offer an "explanation." The goal of explainable ML is to make the decision-making process of ML models more transparent and interpretable, which is especially important in safety-critical domains.

We utilize the "model distillation" [5] for achieving explainable machine learning. This approach involves developing a separate model, called the "distilled model," that approximates the input–output behavior of the target machine learning model. The distilled model is inherently explainable, enabling users to identify the decision rules or input features that influence the final decision, as shown in Fig. 3. This model is commonly developed in a "white-box" manner to reveal the decision rules on how input features influence the ML outputs. Therefore, lightweight structures such as linear regression [31], decision tree [39], or object graph [38] are preferred.

We also investigate Shapley values, borrowed from cooperative game theory [32], to achieve explainable machine learning. Shapley values measure a player's individual contribution to the outcome by capturing the marginal contributions. For the i-th player, the marginal contribution can be calculated formally by

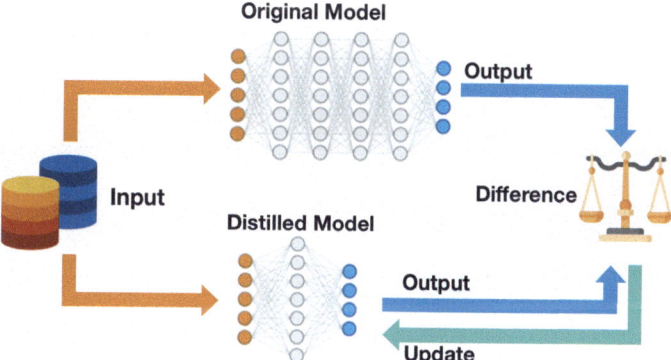

Fig. 3 Model distillation minimizes the difference of input–output mapping behaviors. The distilled model is inherently explainable since it provides insights into the internal representation to explain how it makes a decision

$$\phi_i = \sum_{S \subseteq N/\{i\}} \frac{|S|!(P - |S| - 1)!}{P!} [v_x(S \cup \{i\}) - v_x(S)] \tag{1}$$

Here, the total set of players is denoted as $|P|$. S is a subset of $|P|$ that excludes player i, and $v_x(\cdot)$ calculates the total gain given subset S. Intuitively, Shapley values tell us the payoff gain with the participation of player i.

Explainability is a crucial characteristic for our proposed work, providing two significant advantages over traditional ML-based detection of Spectre and Meltdown attacks. Firstly, our method utilizes explainability to identify critical features, requiring only a small set of crafted samples for adversarial training. Without explainability, extensive training with a large number of samples would be necessary, consuming substantial time and space. Figure 9 illustrates how explainability aids in crafting synthesized samples with patterns similar to evasive samples. Secondly, explainability allows for interpreting predictions in a human-readable way that can handle incorrect classification results.

- Existing approaches consider input data that are discrete values. In security domain, we need to handle input data that are time-sequential records.
- Existing approaches focus on computer vision tasks using CNN. However, CNN model is not suitable for security applications with time-sequential data.

3 Detection of Spectre and Meltdown Attacks

Figure 4 shows an overview of our proposed detection framework using explainable machine learning that satisfies two important requirements.

- *Design Overhead:* Efficient utilization of hardware features with minimal impact on overhead.
- *Detection Robustness:* Effective countermeasures to protect against obfuscation techniques.

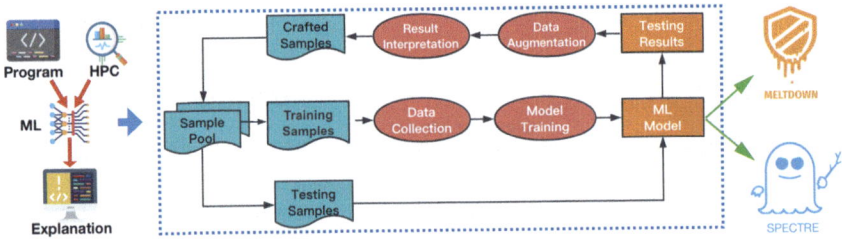

Fig. 4 An overview of Spectre and Meltdown detection framework that consists of four major tasks: data collection, model training, result interpretation, and data augmentation

Figure 4 is an iterative process comprising the following four major tasks. The remainder of this section describes these tasks in detail.

Data Collection To collect hardware event records, we initially execute both benign and malicious programs using HPCs (Sect. 3.1), which serves as our initial sample pool. From this pool, we identify several critical features as important events, which are then selected for the machine learning training process.

Model Training The structure of our ML model is built upon a recurrent neural network, which is trained using stochastic gradient descent. Further details on the training process, including the selection of features, are outlined in Sect. 3.2.

Result Interpretation To test the trained model, we employ a set of test samples to generate classification results, which are then utilized by the *result interpretation* task described in Sect. 3.3. It provides insights into the most relevant (or misleading) input features for classification.

Data Augmentation Building upon the analysis obtained from result interpretation, we craft adversarial samples and integrate them into the original pool of training samples for model retraining, as outlined in Sect. 3.4. The model continually improves itself until testing accuracy converges. This well-trained model is then utilized for automatic attack detection.

3.1 Data Collection

The initial step in training our ML model involves collecting and determining the format of the model inputs. As discussed in Sect. 1.2, we found that utilizing hardware events shows promise in ML-based attack detection. However, it is a major challenge to select a small set of hardware performance counters (HPCs) that are beneficial for attack detection out of a large number of HPCs that monitor a vast array of hardware events in the modern microprocessors. For Meltdown attacks, the out-of-order memory lookup generates a significantly high number of page faults, which serves as an effective indicator. For Spectre attacks, we record the total number of branch instructions and mispredictions to compute branch miss rate since Spectre attacks abuse branch prediction properties. Furthermore, since both attacks rely on a cache-based side-channel attack, we collect the total number of low-level cache (LLC) references and miss to detect suspicious cache events. Based on these observations, we have selected six critical features, as shown in Table 1, to use in training our model.

Crafting vectors composed of the selected features is a straightforward way of formatting the hardware events for training our ML model. This naive strategy has been adopted by state-of-the-art works [2, 18]. For example, Fig. 5a shows the distribution of normal and malicious Spectre attack samples with LLC misses, LLC references, and branch miss rate. It is easy to distinguish between malicious samples and normal ones, with obvious regions and boundaries between the two classes. This

Table 1 Beneficial hardware performance counters for detection of Spectre and Meltdown attacks

Hardware events	Event ID	Spectre	Meltdown
Total number of instructions	INS	✓	✓
Total page faults	PGF	✗	✓
Total branch instruction	BRC	✓	✗
Branch Miss-predictions	BMP	✓	✗
Low-level cache references	LLCR	✓	✓
Low-level cache misses	LLCM	✓	✓

observation confirms that the selected features are beneficial in detecting attacks. However, this naive approach is ineffective against evasive attacks. In evasive attacks, the adversary can add redundant loops or cache access statements, enabling the malicious program to mimic the pattern collected from benign programs, making the overall statistics indistinguishable from normal programs. Figure 5b shows the distribution of evasive Spectre samples, which are mingled with normal ones and have no clear boundary to distinguish them. As demonstrated in Sect. 4, evasive attacks drastically reduce the performance of state-of-the-art detection methods.

To overcome the limitations of the naive approach, we applied two strategies. First, we designed the data format in our approach as an *event-tracing table*, as illustrated in Table 2. Instead of using overall statistics, we sample hardware events using HPCs at multiple timestamps and record their differences. Each row in the table represents a specific selected hardware event, and the "Δ" symbol denotes the increase of the corresponding event compared to the previous timestamp. By considering the hardware events in sequential timestamps, this strategy enables the model to gain valuable information hidden in consecutive adjacent inputs. Next, we employed *data augmentation* during the ML training process, as discussed in Sect. 3.4.

3.2 Model Training

Given the time-sequential nature of our data, we have chosen to use a *Recurrent Neural Network (RNN)* as the main structure of our model. The RNN is represented by M, where $y_0, y_1, y_2, \ldots$ correspond to the time series inputs (columns of the event-tracing table), and r_is are the outputs of the hidden layers. Unlike standard neural networks that perform input–output mapping in one forward pass, the RNN takes in sequential inputs. For each input y_i, the RNN provides an immediate response r_i as well as information from the previous step, supplying additional context. This mapping is visualized through the unrolled RNN structure shown in Fig. 6. To mitigate the *vanishing gradient* and the *exploding gradient* problems outlined by Bengio et al. [3], we employ two different model training approaches. The first method involves implementing the RNN as a *long short-term memory*

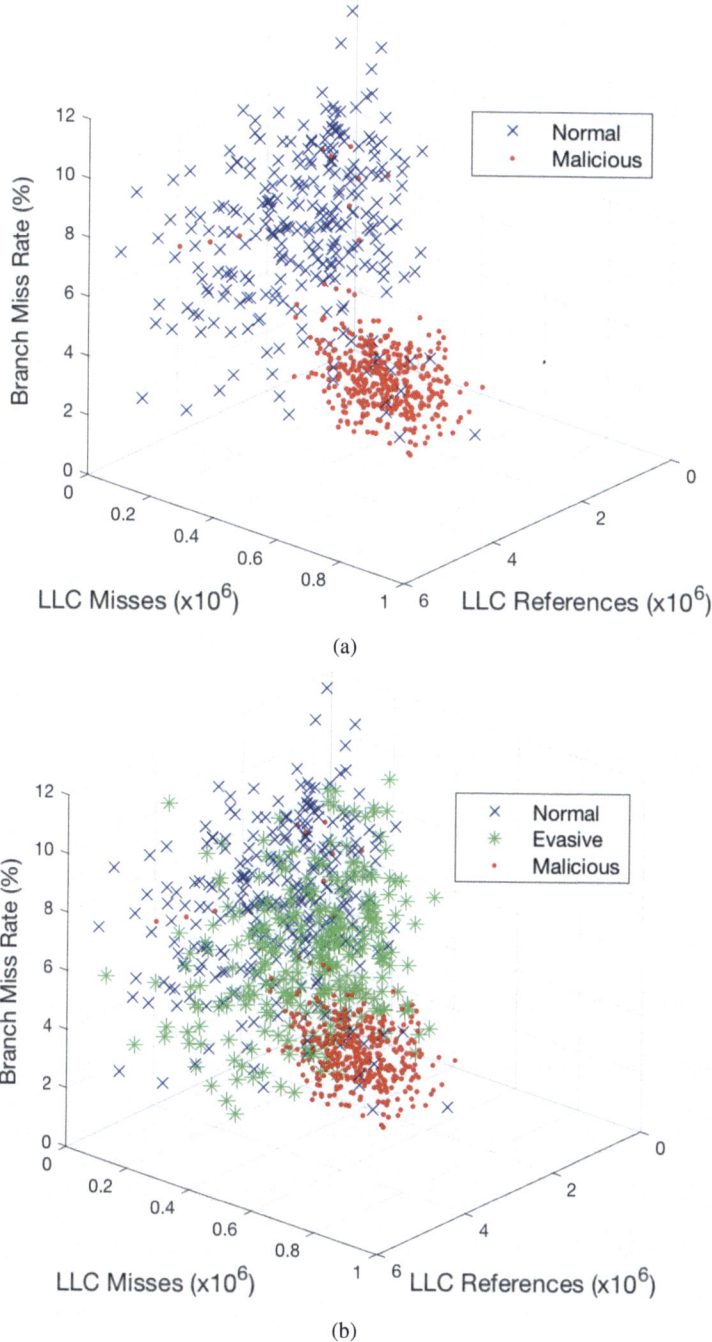

Fig. 5 Distribution of LLC misses, LLC references, and branch miss rate for different types of samples. (**a**) Normal versus malicious. (**b**) Normal versus malicious vs evasive

Table 2 An example record illustrating differences in event traces

Time\Events	y_0	y_1	y_2	y_3
Δ BMP	0.3	0.1	−0.2	...
Δ LLCR	45	73	37	...
Δ LLCM	81	24	69	...
...				

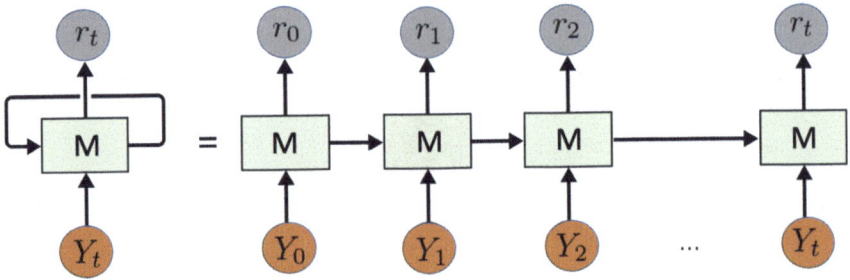

Fig. 6 Recurrent Neural Network (RNN)

(LSTM), which is described in Sect. 3.2.1. The second method utilizes ensemble boosting, as discussed in Sect. 3.2.2.

3.2.1 LSTM-Based Model Training

LSTM is a distinct variant of RNN that incorporates specialized units in addition to the basic configuration. It features a "memory cell" that can retain information in memory over long periods and a set of gates that regulate the information flow inside LSTM's architecture. This setup enables LSTM to learn longer term dependencies. Figure 7 provides an overview of our proposed approach, while Algorithm 1 summarizes the important steps during the training process.

In our approach, the LSTM serves as an auto-encoder to map the input data to hidden features. However, we require an additional structure to map the learned "distributed feature representation" to a more complex feature space. Thus, we employ a *multi-layer perceptron* (MLP) neural network to process the output of the last LSTM layer. The output of the MLP is then normalized by a *softmax* layer, which generates binary prediction labels. To train the model, we use the *cross_entropy* loss function, and its gradient is used to update the model parameters using the *stochastic gradient descent (sgd)* method. An overview of this architecture is shown in Fig. 7. The major steps of the training process are summarized in Algorithm 1.

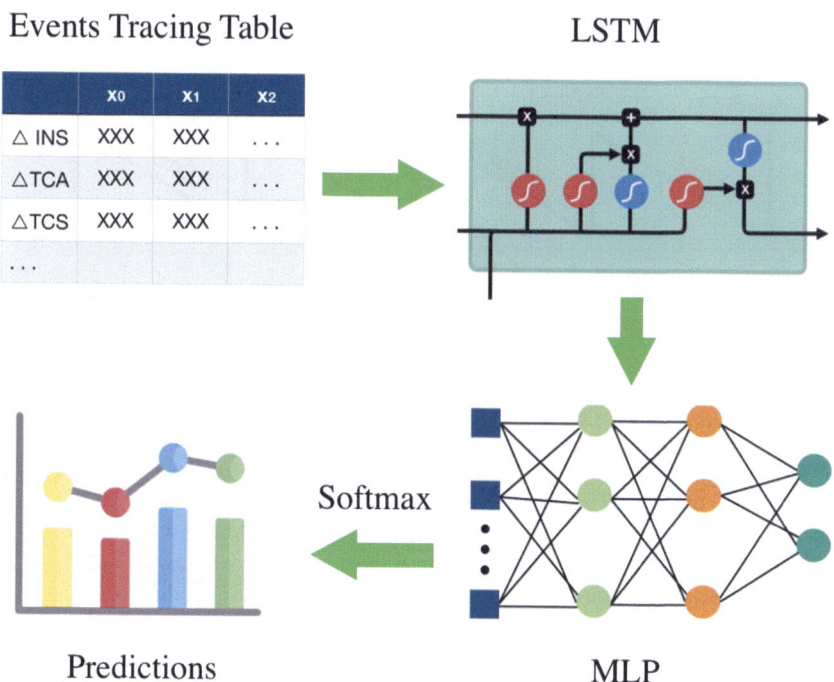

Fig. 7 An overview of the proposed ML model that consists of two major components: LSTM and MLP. The output of the last hidden layer passes through softmax function to produce prediction labels for input samples

---

Algorithm 1: LSTM-based model training

    Input : Model Inputs $\{y_i\}$
    Output: Trained Model M
1 initialize(M)
2 $r_0 = M(y_0)$
3 **repeat**
4   **for** $i = 1 \dots t$ **do**
5        $r_i = M(y_i, r_{i-1})$
6     $res = \text{softmax}(r_t)$
7     $loss = \text{cross\_entropy}(res, label)$
8     $M = \text{sgd}(M, \nabla loss)$
9 **until** $converge$;
10 Return M

3.2.2   Ensemble Boosting

Ensemble boosting is another method of model training that combines multiple weaker sub-models to create a stronger model. Each sub-model is trained in sequence, where every subsequent model focuses on mitigating the errors produced

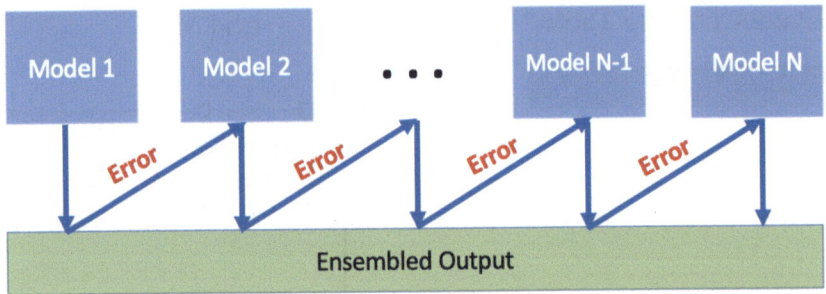

Fig. 8 The structure of the proposed ML model based on boosting. The entire framework is an ensemble model consists of a set of lightweight RNN classifiers. Subsequent models aim at fixing mispredictions caused by previous models, and the final output is the voting result from all models

by previous models. The final prediction is based on the voting result of all the individual sub-models, which leads to faster prediction and higher accuracy without sacrificing generalization. In Fig. 8, we show an overview of the proposed framework. We chose boosting as the second implementation method because it has the potential to improve both detection accuracy and time efficiency of predictive models. In ensemble boosting, we randomly sample a subset of the input dataset for training each sub-model instead of blindly feeding all the data. The key step in this method is the weight adjustment operation, which tunes the possibility of selecting samples in the next iteration of training. The goal is to increase the chance of sampling hard-to-fit samples in future iterations. Additionally, benchmarks with more serious prediction errors (larger loss function output) should be given higher priority to speed up the convergence.

Let $X_i \in \mathcal{D}$ be the i-th data sample from dataset $\mathcal{D}$, where X_i's corresponding label is denoted as y_i. Here, $y_i = 0$ represents a benign program and $y_i = 1$ represents a malicious one. The probability of selecting benchmark X_i is denoted as $\hat{y}_i$, and the weight of X_i is recorded as w_i^b. For a given benchmark X_i, we have

$$\Delta w_i^b \propto \mathcal{L}(y_i, \hat{y}_i)$$

where $\mathcal{L}$ is the loss function (cross entropy), reflecting the extent of prediction error deviated from the ground truth:

$$\mathcal{L}(y_i, \hat{y}_i) \triangleq (y_i \, log \, \hat{y}_i + (1 - y_i)log(1 - \hat{y}_i))$$

which leads to

$$\Delta w_i^b \propto (y_i \, log \, \hat{y}_i + (1 - y_i)log(1 - \hat{y}_i))$$
$$\Rightarrow w_{i_{(t+1)}}^b = w_{i_{(t)}}^b + \alpha(y_i \, log \, \hat{y}_i + (1 - y_i)log(1 - \hat{y}_i))$$

where t is the iteration count and α is the learning rate. We have artificially constrained the step size to be small in our framework to increase the likelihood of the parameters converging to optimal values, thereby improving the accuracy of the model. The complete training process for ensemble boosting is outlined in Algorithm 2.

Algorithm 2: Boosting-based model training

Input   : Dataset($\mathcal{D}$), learning rate (α), feature set ($\mathcal{F}$)
Output: SEB Model $\mathcal{M}$

1 $t = 0$
2 For each $X_i \in \mathcal{D}$, $w_{i_{(t)}}^b = \frac{1}{sizeof(\mathcal{D})}$
3 **repeat**
4 $\quad$ Random Sample $\mathcal{D}' \subset \mathcal{D}$
5 $\quad$ $M_{(t)} = \text{RNN}(\mathcal{D}')$
6 $\quad$ **for** *each $X_i \in \mathcal{D}'$* **do**
7 $\quad\quad$ $\hat{y}_i = M(X_i)$
8 $\quad\quad$ $w_{i_{(t+1)}}^b = w_{i_{(t)}}^b + \alpha \mathcal{L}(y_i, \hat{y}_i)$
9 $\quad$ $t = t + 1$
10 $\quad$ $\mathcal{M} = \mathcal{M} \cup \{M_{(t)}\}$
11 **until** *convergence*;
12 Return $\mathcal{M}$

3.3 Result Interpretation

We have trained a model for the detection activity, but it can still be vulnerable to evasion methods discussed in Sect. 1.2. To address this issue, we leverage explainable machine learning to interpret the detection outcomes, which can be utilized to generate evasive data samples. These samples are added to the training set, effectively enhancing the model's robustness against known attacks. We adopt two different approaches to interpret the results: model distillation and Shapley value analysis. The remainder of this section describes these approaches in detail.

3.3.1 Explainability Using Model Distillation

To achieve result interpretation, we utilize model distillation that consists of three major steps: model specification, model computation, and outcome explanation.

Model Specification To select an appropriate model for distillation, a trade-off must be made between the model's transparency and its expressive ability. While a complex model may yield better performance in mimicking the original model, it often results in reduced model transparency. Conversely, simpler models are more interpretable but may sacrifice performance. We use a linear regression model due to its simplicity and interpretability.

Model Computation Once the type of distilled model ($\mathbf{M}^*$) is determined, the original model $\mathbf{M}$ passes through test samples to produce the required number of input–output pairs. The objective of this task is to search for optimal parameters θ to minimize the difference between $\mathbf{M}$ and $\mathbf{M}^*$. Since we are solving using linear regression, this task is a least square problem and can be solved efficiently as

$$\theta = \arg\min_{\theta} \sum_{i=1}^{n} ||\mathbf{M}^*_{\theta}(\boldsymbol{y}_i) - \mathbf{M}(\boldsymbol{y}_i)||_2 \tag{2}$$

where vector $\boldsymbol{y}_i = [y_{i1}, y_{i2}, \ldots]$ is the i-th input.

Outcome Explanation To interpret the distilled model, we need to determine the contribution of each input feature in producing the model output. In case of linear regression, the model can always be expressed as a polynomial, allowing us to sort the terms with the amplitude of their coefficients and gain insights into the input features that are most discriminatory. For example, if the distilled model after regression is $\mathbf{M}^*\boldsymbol{\theta}(yi) = a_1 y_{i1} + a_2 y_{i2} + a_3 y_{i3} + \ldots$, we can sort the terms by the absolute value of their coefficients. If a_j is the largest coefficient in the term $a_j y_{ij}$, then the most important contributor is y_{ij}. This process provides a transparent and interpretable explanation of the model's behavior, allowing for improved trust and understanding of the model's decision-making process.

In this task, the machine learning model is able to identify the most significant elements of each input $\boldsymbol{x}_i$ by listing and sorting the contributions of the input features. This represents the entries of the i-th column from the event tracing table that are main contributors to the model output. This also indicates the hardware events that occurred in the i-th timestamp that were considered by the machine learning model. This is particularly important for improving the model's performance in handling incorrect classifications. By identifying the critical location that led to misprediction, the model can be refined. In Sect. 3.4, we describe how we craft synthesized samples based on these observations.

3.3.2 Explainability Using Shapely Values

Shapley value analysis is a powerful tool for interpreting machine learning (ML) models, particularly when it comes to explaining the role of individual features in making predictions. In this approach, we think of the features as "players" in a cooperative game and use Shapley values to assign credit to each feature based on its contribution to the model's overall performance. To apply Shapley value analysis to a specific ML task, we need to select a set of features that are relevant to the problem at hand. For example, suppose we are building a model to detect Spectre

Table 3 The marginal
contributions of feature 1

Features	Contribution of feature 1
1,2,3	$\mathcal{V}(\{1\}) - \mathcal{V}(\emptyset)$
1,3,2	$\mathcal{V}(\{1\}) - \mathcal{V}(\emptyset)$
2,1,3	$\mathcal{V}(\{1, 2\}) - \mathcal{V}(\{2\})$
2,3,1	$\mathcal{V}(\{1, 2, 3\}) - \mathcal{V}(\{2, 3\})$
3,1,2	$\mathcal{V}(\{1, 3\}) - \mathcal{V}(\{3\})$
3,2,1	$\mathcal{V}(\{1, 2, 3\}) - \mathcal{V}(\{3, 2\})$

and Meltdown vulnerabilities, and we have identified three features that may be useful in this task. To compute the Shapley values for these features, we start by creating a null model with no independent features. We then gradually add each feature to the model in sequence and compute the payoff gain at each step. Finally, we take the average over all possible sequences of feature additions. In our example with three independent variables, we need to consider six different sequences of feature additions. To illustrate the computation process for the Shapley values of one specific feature, we present a table (Table 3) that shows how output values change with sub-sequential features added to the model. This approach provides an intuitive way to understand how each feature contributes to the overall performance of the ML model.

The value function $\mathcal{V}$ is the scoring metric to quantify the current payoff obtained through the application of existing features. Specifically, the first row of the sequence $(1, 2, 3)$ represents the sequential addition of features 1, 2, and 3 to the given model. The absence of any features is denoted by the symbol $\emptyset$, which in the scenario of classification is a random guesser, and its corresponding loss is denoted by $\mathcal{V}(\emptyset)$. By introducing the first feature, we ended up with an initial model that employs solely feature 1 for prediction, represented by 1. The score associated with this model is calculated as $\mathcal{V}(1)$. The marginal contribution is computed as the difference between the current loss values with the starting random guesser, namely $\mathcal{V}(1) - \mathcal{V}(\emptyset)$. The Shapley values for the first feature are subsequently obtained by computing the marginal contributions for all six permutations. Similar processes are performed on other features and we take their average to be the Shapley value.

3.4 Data Augmentation

Upon interpretation of the model's output, valuable insights can be gained from incorrect predictions, allowing for the identification of areas where the model can be improved. To prevent similar errors from occurring in the future, a popular technique known as *data augmentation* can be employed. This technique involves the generation of new, synthesized samples based on existing malicious samples, thereby expanding the dataset and allowing the model to learn from a more diverse range of inputs. The data synthesis process typically comprises five major steps.

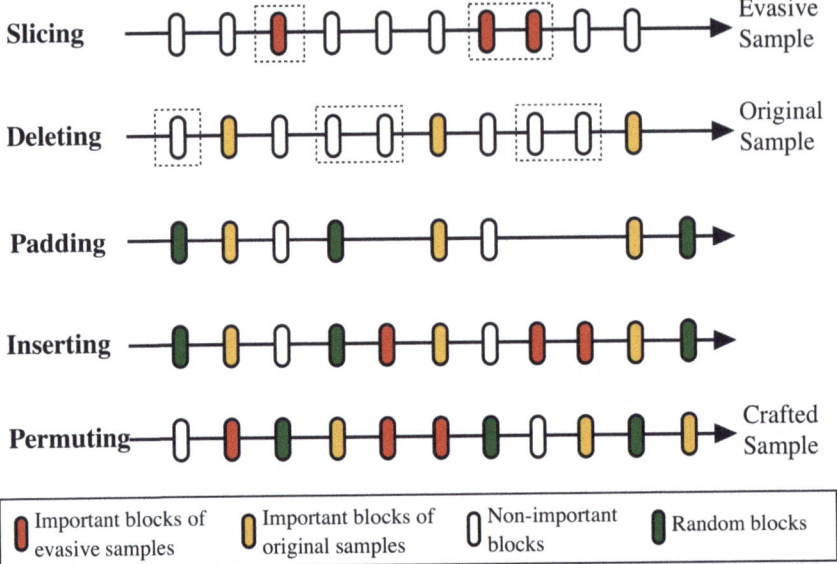

Fig. 9 Illustrative examples of data augmentation. Each row is a small fragment of the full execution log. Each box represents one basic function block. Crafting synthesized sample consists of five stages. First, important misleading activities were sliced from evasive attack codes. Next, non-important blocks from target sample are randomly deleted to prune the size, followed by padding of non-profitable blocks. Then, the misleading blocks are inserted into target sample. Finally, permutation of basic blocks completes the synthesis of new evasive samples

- **Slicing:** With the help of explainable ML, it starts by marking and slicing codes corresponding to the most important timesteps from evasive samples that are incorrectly labeled by the classifier. These code slices are denoted as "inducements."
- **Deleting:** For the original sample, randomly delete non-important and irrelevant parts to prune the size.
- **Padding:** Non-profitable code slices were randomly generated and augmented in the sample to mess-up the overall statistics.
- **Inserting:** Bootstrapping method was applied to sample the pool of "inducements" and to insert them into the target sample.
- **Permuting:** Function blocks are permuted to reorder hardware events, making the fluctuation of HPC records different from original samples.

Figure 9 illustrates the five steps of data augmentation. Algorithm 3 shows how data augmentation is used to retrain the model.

Algorithm 3: Model training with data augmentation

Input : Initial model (**M**), distilled model (**M***), sample pool S, number of iterations (t)
Output: Optimized model, **M**$'$

1   $i = 1$
2   **repeat**
3     $S' = \{y \in S | \mathbf{M}(y) \neq label(y)\}$
4     $rank = sort(\mathbf{M}^*.\mathbf{coeff})$
5     $S_{adv} = craft(S', rank)$ ▷ Craft adversarial samples
6     $S = S \cup S_{adv}$
7     $\mathbf{M} = train(\mathbf{M}, S)$ ▷ Model retraining
8     $\mathbf{M}^* = explainableAI(\mathbf{M}, S)$
9   ▷ Utilize model distillation or Shapley analysis
10   i++
11 **until** $i \geq t$;
12 Return **M**$'$

---

4 Experiments

In this section, we demonstrate the effectiveness of our Spectre and Meltdown detection framework. We first outline the experimental setup used in our evaluation. Next, we present the Spectre and Meltdown detection results.

4.1 Experimental Setup

We conducted our experimental evaluation on a host machine with an Intel i7 3.70GHz CPU, 32 GB RAM, and RTX 2080 256-bit GPU. Our implementation was developed using Python. We utilized PyTorch as the machine learning library.

To ensure a fair comparison with related methods, we conducted our experiments using the following four sets of subject programs:

- The purpose-built code examples to demonstrate different variations of Spectre vulnerabilities from Paul Kocher's blog post [15].
- The Standard Performance Evaluation Corporation (SPEC) benchmarks that consist of two major branches. SPECint (integer) focuses on vulnerability of control-intensive codes. SPECfp (floating point) mainly exploits vulnerability through conditional branches [11].
- Google's OSS Fuzz repository posted on GitHub [17] covering commonly seen problematic software.
- Virtual Machine (VM)-based meltdown attack samples from Seed Lab [1].

In addition to incorporating the original dataset, we also manually crafted evasive attack samples using the following two obfuscation techniques as introduced in [19].

1. **Strategy 1:** Put attack into sleep between memory flush.

2. **Strategy 2:** Insert redundant instructions for obfuscation.

As a result, the total dataset consists of 1000 samples to enable a comprehensive evaluation. Based on that, the performance counter values are collected at a sampling rate of 100ms during program execution, and the collected traces are formatted into an event-tracing table, as shown in Table 2. Our training dataset includes data collected from 200 runs of both malicious and benign programs, resulting in an average data size of approximately 19.2 KB per run. Thus, the total data size for each program is 3840 KB (200 x 19.2). To ensure independence across different runs, the system status was reset after each run, and a realistic scenario for data collection was considered.

In terms of the proposed ML model architecture, as outlined in Sect. 3, we implemented our proposed approach in two different ways. The first model comprises a Long Short-Term Memory (LSTM) network and a Multilayer Perceptron (MLP). The LSTM architecture comprises a one-hot encoding layer, a hidden layer with 32 nodes, and a 50% dropout. The MLP comprises three layers and 64 nodes. The second model adopts the boosting framework idea and consists of 100 boosting stages with a sample rate of 0.5 and the learning rate set as default to 0.1. Our input data consists of HPC values collected during the execution of both malicious (with implanted Spectre/Meltdown attack) and benign programs.

Based on the above configuration, we trained our proposed explainable defense technique with two variants: (i) LSTM with model distillation (**Proposed LSTM**) and (ii) Shapley ensemble boosting (**Proposed Boosting**). While LSTM and boosting enhance training efficiency, model distillation and Shapley analysis provide improved explainability. The effectiveness of our proposed method is evaluated by comparing it with several state-of-the-art defense techniques. As discussed in Sect. 2.1, existing defense strategies can be categorized into two types, detection and mitigation. We summarize existing state-of-the-art defense strategies in Table 4 and compare the performance with our proposed method.

Table 4 State-of-the-art detection and mitigation of Spectre and Meltdown attacks

Methods	Description
Detection algorithms	(i) **RDSM:** State-of-the-art detection framework for both Spectre and Meltdown attacks [2]
	(ii) **AT-RDSM:** We extended RDSM by training with adversarial samples to enable fair comparison
	(iii) **ODSA:** Efficient online detection approach for Spectre attack using various implementations [18]
	(iv) **oo7:** State-of-the-art low-overhead defense against Spectre attacks via program analysis [34]
Mitigation techniques	(i) **KPTI:** Kernel page-table isolation, a commercial solution to mitigate Meltdown using Linux kernel features [25]
	(ii) **S-Mitigation:** A combinational technique including Indirect Branch Restricted Speculation (IBRS) and Return Stack Buffer (RSB) refilling [9]

4.2 Comparison with Existing Spectre Detection Methods

The ODSA framework [18] includes five different implementations for detecting Spectre attacks, which are summarized in Table 6 in terms of their effectiveness. Each implementation is evaluated based on its detection rate (DR), false positive (FP) rate, and false negative (FN) rate. Based on the evaluation results, MLP showed the best performance among the ODSA implementations. Therefore, we compare our approach with the MLP implementation in Table 5.

Table 5 presents a performance comparison of our proposed approach with state-of-the-art methods for detecting both Spectre and evasive Spectre attacks. Among the ODSA implementations, MLP provides the best detection rate for normal Spectre attacks, but its performance drops significantly to around 50% when detecting evasive attacks, making it unsuitable for deployment. This pattern of performance is more obvious when we looked at oo7, which achieves nearly perfect detection accuracy for normal attacks (99.2%) but drops down to 53.2% when facing with evasive Spectre attacks. The failure of ODSA and oo7 against evasive attacks is likely due to the similarity of features extracted from evasive samples and benign programs, as shown in the feature analysis in Fig. 5, where linear classifiers such as LR, SVM, or MLP are unlikely to succeed when features are mixed in space. In contrast, our proposed approach significantly outperforms both RDSM and AT-RDSM in detecting Spectre attacks, with an accuracy of 96.2% compared to ODSA's average of 96.7% (as shown in Table 6). Additionally, our approach based on a single LSTM model outperforms ODSA by 28.7% in detecting evasive Spectre attacks, and the boosting-based method further improves the detection accuracy, achieving 98.5% accuracy for normal cases and 89.9% accuracy against adversarial Spectre attacks.

To showcase the efficacy of our proposed explainable framework, we present the distribution of misclassifications for four categories in Fig. 10. The false positive (FP) category includes the scenario when a benign program is classified as malicious. The false negative (FN) category refers to the scenario when a malicious program is classified as benign. We further divide the false negative cases into three subcategories: evasive attacks, and two obfuscation strategies (Strategy 1 and Strategy 2, as outlined in Sect. 4.1). We observe that ODSA and oo7 are highly susceptible to Strategy 2, where adding redundant instructions confuses these two approaches and results in false negative predictions. This is in line with the fact that they are based on program analysis mechanism, where program features and hardware events (e.g., LLC miss rate and branch miss rate) are used as the primary measurements for classification. Unfortunately, these hardware events can be easily manipulated by an adversary. Therefore, Strategy 1 has little impact on misclassifications since randomly putting the program to sleep will not affect the above events. However, the branch selection and cache references introduced by redundant execution are detrimental to ODSA and oo7.

Table 5 Comparison of Spectre attack detection methods

Attack type	RDSM [2]			AT-RDSM			ODSA-MLP [18]			oo7 [34]			Proposed LSTM			Proposed boosting			
	DR (%)	FP (%)	FN (%)	DR (%)	FP (%)	FN (%)	DR (%)	FP (%)	FN (%)	DR (%)	FP (%)	FN (%)	DR (%)	FP (%)	FN (%)	DR (%)	FP (%)	FN (%)	Impr/ oo7(%)
Normal	89.1	7.7	3.2	94.1	4.3	1.6	99.2	0.8	0.0	99.8	0.1	0.0	96.2	2.4	1.4	98.5	1.0	0.5	−1.3
Evasive	22.1	39.3	38.6	58.4	27.5	14.1	59.4	20.4	20.2	53.2	26.3	20.5	88.1	4.4	7.5	89.9	5.2	4.9	36.7
Average	55.6	23.5	20.9	76.2	15.9	7.9	79.3	10.6	10.1	76.5	13.2	10.3	92.2	3.4	4.4	94.2	3.1	2.7	17.7

Table 6 Different ODSA
implementations [18]

Implementations	DR (%)	FP (%)	FN (%)
LR	92.8	3.8	3.4
Tuned LR	96.4	1.1	2.5
SVM	96.8	0.7	2.5
Kernel SVM	98.3	0.7	1.0
MLP	99.2	0.8	0
Average	96.7	1.4	1.9

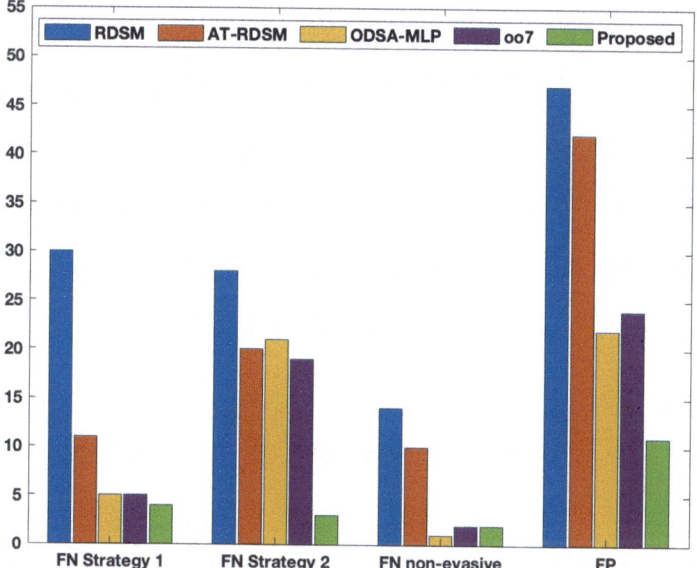

Fig. 10 Distribution of incorrect Spectre classification for both false positive (FP) and false negative (FN) results

4.3   Comparison with Existing Meltdown Detection Methods

We compared the performance of our proposed method with RDSM and AT-RDSM for detecting Meltdown attacks, as shown in Table 7. Since ODSA did not report any results for Meltdown attacks, we only considered RDSM and AT-RDSM. We also included the improvement achieved by our approach over AT-RDSM in the last column. While AT-RDSM performed well in detecting non-evasive Meltdown attacks with a detection rate of 95.9%, its performance against evasive Meltdown attacks dropped to 55.6%, which is equivalent to random guessing. This significant drop in performance demonstrates the instability of these methods with respect to obfuscation techniques. In contrast, our approach achieved a high detection rate of 99% for non-evasive attacks in both implementations and still maintained a high accuracy for evasive ones. Moreover, our proposed method outperforms RDSM and

Table 7 Comparison of Meltdown attack detection methods

Attack type	RDSM [2]			AT-RDSM			Proposed LSTM			Proposed Boosting			
	DR (%)	FP (%)	FN (%)	DR (%)	FP (%)	FN (%)	DR (%)	FP (%)	FN (%)	DR (%)	FP (%)	FN (%)	Impr/ AT-RDSM (%)
Normal	93.5	2.9	3.6	95.9	2.5	1.66	99.0	0	1.0	99.0	0.5	0.5	3.1
Evasive	19.2	25.6	55.2	55.6	22.8	21.6	94.5	2.3	3.2	95.5	2.4	2.1	39.9
Average	56.4	14.2	29.4	75.9	12.7	11.4	96.7	1.2	2.1	97.2	1.4	1.4	21.3

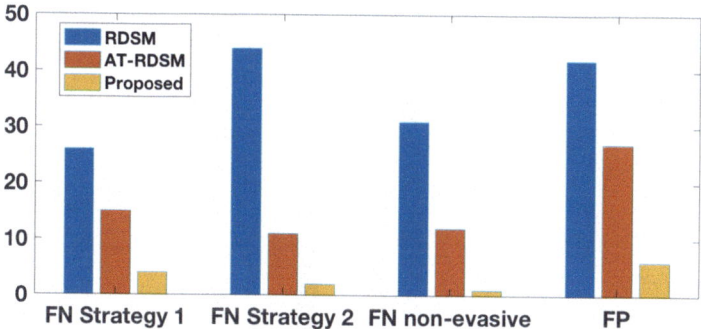

Fig. 11 Distribution of incorrect Meltdown classification results for both false positive and false negative results

AT-RDSM in detecting both evasive Spectre and evasive Meltdown attacks, making it a robust and reliable solution.

Table 7 highlights the limitations of previous methods in terms of generating high false positive results. In Fig. 11, we demonstrate the distribution of all misclassified inputs for four different categories. It is evident that RDSM is highly vulnerable to obfuscation Strategy 2, which can obscure the patterns of malicious behavior to the extent that a classifier is unable to distinguish between malicious attacks and benign programs. Although AT-RDSM improves the robustness against evasive attacks to some extent, it still has a high false positive rate due to its inability to interpret the reason for misclassification. The lack of interpretation means that users do not know the exact reason for the incorrect prediction and blindly feed adversarial samples. This can lead to "overfitting" problem since the model may learn some benign features in these samples and induce a high false positive rate.

Based on the above discussion, we have identified three major advantages of our proposed approach over state-of-the-art methods:

- *Alternative Structure:* Our approach uses an RNN, which is capable of handling time-sequential data and making decisions based on potential information hidden in consecutive adjacent inputs. This results in more reliable classification outcomes.
- *Explainable Interpretation:* One critical difference between our approach and AT-RDSM is that we perform outcome interpretation before adversarial training, enabling us to craft adversarial samples more effectively.
- *Ensemble Aggregation:* In both Spectre and Meltdown detection tasks, our boosting-based implementation outperforms the single LSTM-based model. Boosting involves sequentially training weak learners, with each learner attempting to rectify the mistakes of the previous one. This significantly improves overall prediction accuracy.

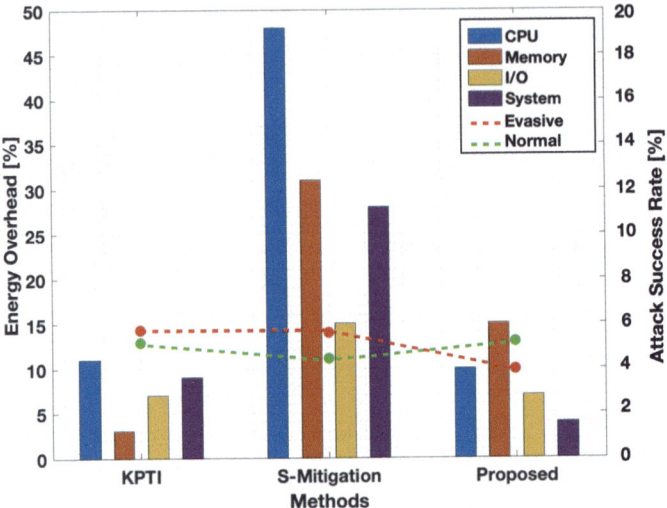

Fig. 12 Comparison of the energy overhead of proposed approach and various mitigation techniques, along with their performance against both normal and evasive attacks

4.4 Comparison with Existing Mitigation Techniques

Mitigation is a widely adopted defense strategy against Spectre and Meltdown attack. Instead of detecting the occurrence of malicious attack, mitigation aims at alleviating the consequence or preventing attacks from happening, which commonly involves memory refresh operations or specifically designed micro-architecture. Generally speaking, mitigation techniques provide promising defense performance, but in a price of higher energy overhead. In Fig. 12 we present the comparison of the overhead of proposed approach and various mitigation techniques, along with their performance against both normal and evasive attacks.

In this figure, we provide the bar plot of the CPU, memory, I/O, and system overhead for KPTI [25], S-Mitigation [9], and the proposed method, respectively. Meanwhile, we plot the Attack Success Rates (ASRs) in dotted lines of both normal and evasive attacks against all the defense strategies. As we can observe from the figure, our proposed approach consumes significantly less energy than S-Mitigation. This is expected since our proposed method does not involve frequent cache refreshing, nor introducing any new hardware component in the system, but merely rely on the feature analysis of HPC values. In contrast, S-Mitigation engages multiple rounds of memory flushing to handle different variants of Spectre attacks. Comparing with KPTI, our approach still induce less energy overhead in terms of CPU, I/O, and system but only causes slightly more memory. We consider this extra cost of memory is caused by the storage of our ML model. In terms of attack success rate, mitigation techniques does not identify the malicious behavior but only focus on alleviating the memory leakage. Therefore, the evasive strategies in program level no longer circumvent the defense, where promising defending performance ($< 15\%$ ASRs) is achieved by all three approaches. Based on the above discussion, we claim that our proposed method is capable of providing comparable defense performance as the state-of-the-art commercial mitigation techniques, with significantly less energy overhead.

4.5 Stability Analysis

For machine learning-based approach, the generalizability and stability is another important concern. During our experiments, we split the entire dataset into three subsets: training, validation, and testing dataset. Specifically, we apply K-fold cross-validation to further improve the model's generalizability as well as avoiding overfitting problem. In our experiment, we choose $K = 10$, and the cross-validation was conducted on both LSTM-based model and Boosting-based model in different datasets for 10 repetitions, respectively. The validation performance is depicted in Fig. 13, where we present the box-plot of prediction accuracy for different models on Spectre and Meltdown validation subset within 10 repetitions. Figures in the first row corresponds to the performance on Spectre attack samples, while the second row corresponds to Meltdown attack, with the columns related to LTSM-based and boosting-based model, respectively. The gray triangles in the diagram indicate the mean value, and the circles represent the outliners. As we can see from the figure, both proposed models achieve $> 90\%$ average accuracy during validation stage. LSTM-based models possess less stability compared with Boosting-based models but tend to have less outliers. This is expected, since the prediction error occurred during training stage might be exaggerated by LSTM's recurrent structure, causing larger variance of performance. While boosting builds each sub-model on previous trees' residuals/errors, the occurrence of outliers will have much larger residuals than non-outliers, so during the training stage, the aggregated architecture is highly likely to pay a disproportionate amount of attention on those points by generating another outlier in opposite direction to balance its effect, which indeed doubles the total number of outliers.

4.6 Explainability Analysis

In this section, we demonstrate the explainability of our proposed Spectre and Meltdown detection approach. We start by testing the explainability of our model through model distillation. In this case, we distilled the original model into a linear model and examined the coefficients learned for each feature to explain their contribution towards the model's output. Figure 14 presents the coefficients of all six features in the distilled linear model for a Spectre attack sample. This plot provides us with meaningful explanations for the model results. For instance, we can

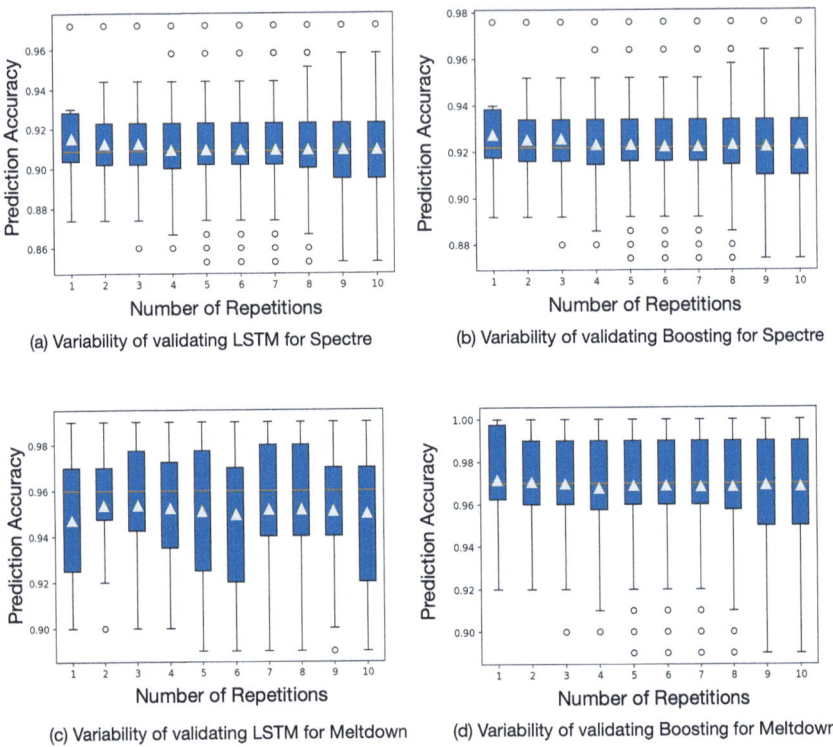

(a) Variability of validating LSTM for Spectre

(b) Variability of validating Boosting for Spectre

(c) Variability of validating LSTM for Meltdown

(d) Variability of validating Boosting for Meltdown

Fig. 13 The box-plot of prediction accuracy for different models on Spectre and Meltdown validation subset within 10 repetitions

observe that PGF and BMP tend to be the most important features for classification since they have larger coefficient values. However, we need to be cautious while interpreting the coefficients alone as they are dependent on the scale of the input features. For example, in this example, the coefficients for INS and BRC are smaller than the others due to their significantly larger values, which results in their smaller coefficients in the linear regression polynomial. Therefore, we cannot solely rely on the magnitude of a coefficient to determine a feature's importance in an ML model.

We first test the explainability by model distillation. In this case, we distill the original model into a linear model, and then we examine the coefficients learned for each feature to explain the contribution of each feature toward the model's output. We plot the coefficients of all 6 features in the distilled linear model for a Spectre attack sample in Fig. 14. This plot can provide us with certain meaningful explanation for model results. For example, the PGF and BMP tend to be the most important features for classification since they have larger coefficient values. However, the coefficients by themselves are not an accurate way to measure the overall importance. This is because the value of each coefficient depends on the scale of the input features. Notice in this example the coefficients for the total

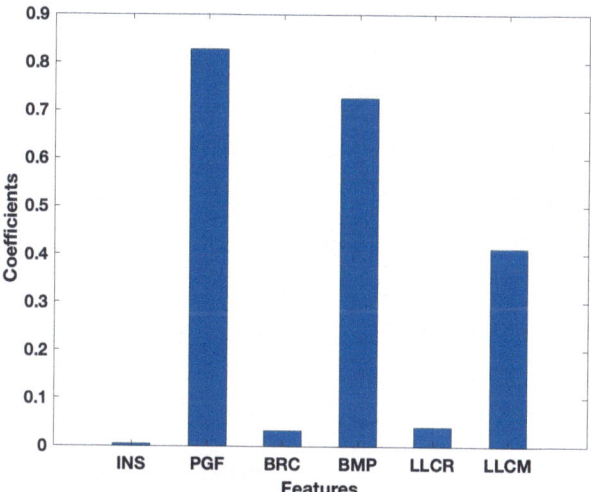

Fig. 14 Histogram for coefficients of all six features in distilled model

instruction numbers (INS) and the total branch instructions (BRC) are far smaller than the others. This is due to the fact that the value of INS and BRC is significantly larger than that of the others. As a result, in the linear regression polynomial, they will be assigned smaller coefficients. Therefore, the magnitude of a coefficient is not necessarily a good measure of a feature's importance in an ML model.

After considering the limitations of the model distillation approach, we explore the explainability of the Shapley value analysis. In machine learning, the classification task typically involves computing a separator and determining which side of the separator it falls on. In the case of Spectre or Meltdown detection, the problem is further simplified to setting a threshold (usually normalized to 0) and comparing it with the model's output. If the output is on the left side (smaller than the threshold), the sample is classified as negative (benign), and if it is on the right side (greater than the threshold), it is classified as positive (attack detected).

The explainability of our proposed Spectre and Meltdown detection approach is tested using the Shapley value analysis, given the limitations of model distillation approach. The classification task in machine learning involves computing a separator and checking which side a sample falls in. In the case of Spectre or Meltdown detection, which is a binary classification problem, the task is further reduced to setting up a threshold, commonly normalized to be 0, and comparing it with the model output. If the output is negative (benign) and falls on the left (smaller) side of the threshold, the sample is classified as negative. Conversely, if the output is positive (attack detected) and falls on the right (larger) side of the threshold, the sample is classified as positive.

Figure 15 plots Shapley values from three samples: (a) and (b) are true positive samples, while (c) is a true negative sample. This waterfall diagram demonstrates the contribution of every single feature selected from Table 1. The plus/minus sign

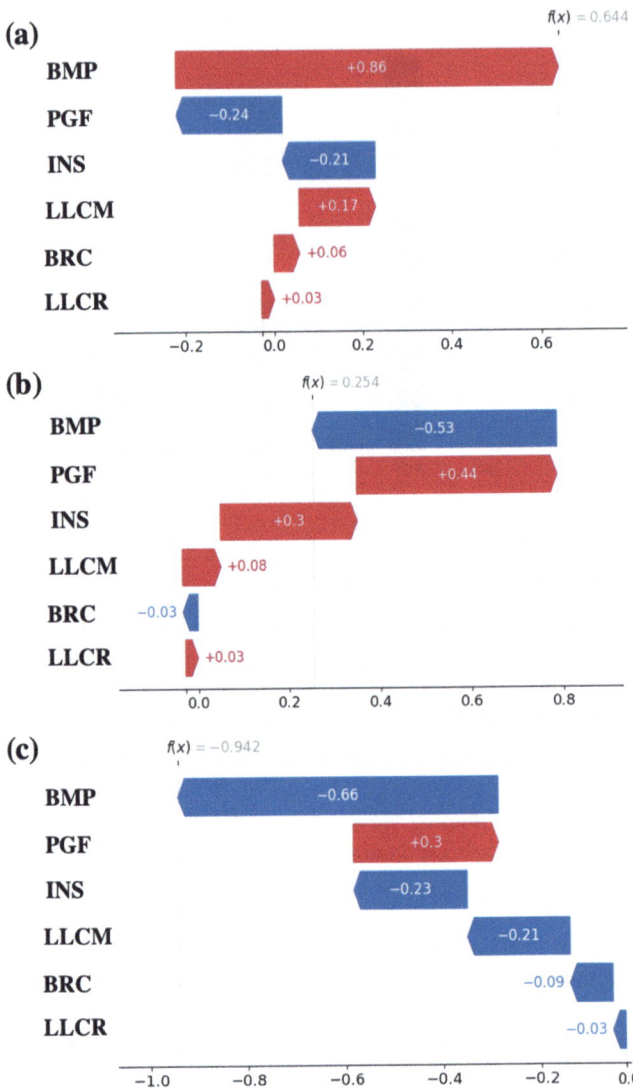

Fig. 15 The SHAP values for three execution scenarios: (**a**) a Spectre attack program, (**b**) a Meltdown attack program, and (**c**) a benign program. The SHAP values clearly illustrate the major features that lead to the model output

indicates how each feature votes for the prediction result, whether it is positive (red bars) or negative (blue bars). The absolute values indicate the amount of impact, with the summation to be compared with the threshold to decide the ensembled prediction. For example, in Fig. 15a, it starts from the threshold (0.0) and the feature "LLCR" pulls the outcome value toward the positive direction (by +0.03), which implies "suspicious." The next feature, "BRC," starts where the previous feature ended and pulls the decision toward the positive direction (by +0.06). This process continues until all the features are taken into consideration. As we can see, the last feature, "BMP," pulls it up to 0.644, which is to the right of the threshold (0.0), and therefore, the final decision is "Spectre attack detected." Similarly in case of Fig. 15b, the last feature, "BMP," pulls it up to 0.254, which is to the right of the threshold (0.0), and therefore, the final decision is "Meltdown attack detected." Finally in case of Fig. 15c, the last feature, "BMP," pulls it up to −0.942, which is to the left of the threshold (0.0), and therefore, the final decision is "benign."

From the figure, we observe that BMP and PGF are among the most important features, which is reasonable and matches our analysis in Sect. 2. In both (a) and (b), the selected attack programs are adversarial samples. In (a), we intentionally caused more page faults to mess up the feature pattern, as the PGF feature provides a negative contribution to the final decision. Nevertheless, the waterfall plot clearly illustrates that our proposed model is still able to assign larger weights to the BMP feature so that it can correctly predict the attack. In (b), we inserted redundant non-profit loops to create more branch mispredictions. The redundant branch mispredictions introduce the biggest negative contribution for the BMP feature to the model's prediction. Since the total number of instructions also increases, the proposed model can produce the correct prediction with the help of the clue from the total instruction numbers, reflected by the positive contribution from INS.

For the purpose of providing a more comprehensible visual representation, we created a decision plot (Fig. 16) to demonstrate the decision-making process based on 10 randomly selected samples, including five positives and five negatives, from the dataset. Each polyline in the plot represents a single sample. The x-axis displays the model's output value, and the plot is centered at 0. The y-axis presents the model's features in descending order of importance. For each polyline, traversing from the bottom to the top provides a complete illustration for the model's decision-making process. At the beginning, every sample starts at the threshold value, and moving upward, the Shapley values are added all the way along. Each polyline will eventually intersect the ceiling at a specific output value, which determines the prediction result. In this figure, out of the 10 lines, 5 are on the left side (negatives) of the central gray line, and the remaining 5 are on the right (positives). Moreover, some polylines' values deviate slightly from 0 even after traversing the bottom five features, indicating that the inclusion of the top feature is crucial for the lines in two classes to start diverging. This finding aligns with our previous discussion.

4.7 Efficiency Analysis

Table 8 compares the time efficiency of four different methods: RDSM, ODSA, LSTM, and Boosting. The table shows the name of each method in the first row, followed by the average training time and testing time in the next two rows.

The boosting framework is the most efficient method across all benchmarks. In contrast, the LSTM-based model lags far behind the others in time efficiency

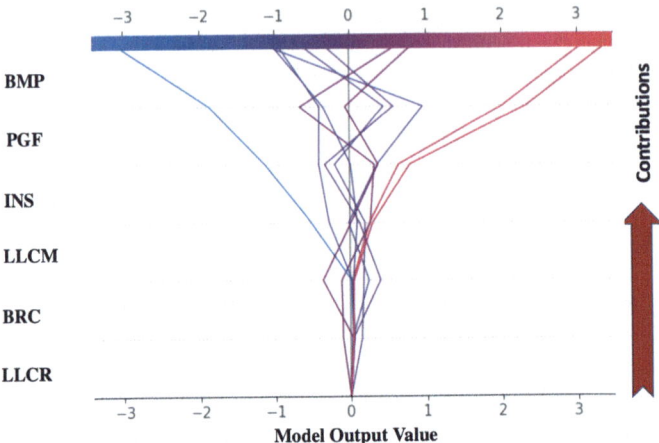

Fig. 16 The decision plot for 10 random samples

Table 8 Comparison of training and testing time (in seconds)

	RDSM	ODSA	Proposed LSTM	Proposed boosting
Training	5831	4791	7724	899
Testing	484	1552	1562	287
Total	6315	6343	9286	1186

due to its complex structure, which requires extensive debugging and parameter tuning. When compared with ODSA and RDSM, the proposed boosting-based method is significantly faster, achieving up to 5.3x speedup. Boosting offers two key advantages in terms of time efficiency. First, unlike the other methods, which dump all data samples and collected features into the training phase, boosting performs partial sampling before training. Second, boosting trains a sequence of lightweight models that require much less time for processing and generate aggregate predictions. The training time of each individual sub-model in boosting is much shorter than that of all the other methods.

5 Summary

Spectre and Meltdown vulnerabilities coupled with the cache-based side-channel attacks are serious threats to modern computer systems. While existing defense mechanisms provide promising results, they have serious limitations including significant performance penalty and hardware overhead. Recently the proposed machine learning-based solutions are also not effective in the face of evasive attacks with obfuscation or other deviation capabilities. In this chapter, we described an explainable machine learning-based detection framework. Our machine learning model is able to make decisions utilizing hardware events generated from hardware performance counters. Moreover, the proposed approach is also able to find the major contributors among all input features to help interpret the classification results, which is further utilized to defend against obfuscation techniques through adversarial training. Experimental results demonstrated that our approach provides comparable performance in detecting Spectre and Meltdown attacks, while it achieves drastic improvement (28.7% for evasive Spectre and 38.9% for evasive Meltdown) in defending evasive attacks compared to state-of-the-art approaches.

References

1. Ramesh Adhikari. Seed-lab-2.0-projects. https://github.com/ramesh-adhikari/seed-lab-2.0-projects, 2017.
2. Bilal Ahmad. Real time detection of Spectre and Meltdown attacks using machine learning. *CoRR abs/2006.01442*, 2020.
3. Yoshua Bengio, Patrice Simard, and Paolo Frasconi. Learning long-term dependencies with gradient descent is difficult. *IEEE transactions on neural networks*, 5(2):157–166, 1994.
4. Claudio Canella, Jo Van Bulck, Michael Schwarz, Moritz Lipp, Benjamin Von Berg, Philipp Ortner, Frank Piessens, Dmitry Evtyushkin, and Daniel Gruss. A systematic evaluation of transient execution attacks and defenses. In *USENIX Security Symposium*, pages 249–266, 2019.
5. Nicholas Carlini and David Wagner. Towards evaluating the robustness of neural networks. In *Security & Privacy*, pages 39–57, 2017.
6. Subodha Charles and Prabhat Mishra. A survey of network-on-chip security attacks and countermeasures. *ACM Computing Surveys (CSUR)*, 54(5):1–36, 2021.
7. Thomas M. Conte, Erik P. DeBenedictis, et al. Rebooting computers to avoid Meltdown and Spectre. *Computer*, 51(4):74–77, 2018.
8. Farimah Farahmandi, Yuanwen Huang, and Prabhat Mishra. *System-on-Chip Security: Validation and Verification*. Springer Nature, 2019.
9. Benedict Herzog, Stefan Reif, Julian Preis, Wolfgang Schröder-Preikschat, and Timo Hönig. The price of Meltdown and Spectre: Energy overhead of mitigations at operating system level. In *Proceedings of the 14th European Workshop on Systems Security*, pages 8–14, 2021.
10. Yuanwen Huang, Swarup Bhunia, and Prabhat Mishra. Scalable test generation for trojan detection using side channel analysis. *IEEE Transactions on Information Forensics and Security (TIFS)*, 13(11):2746–2760, 2018.
11. Krishna Kant. *Introduction to computer system performance evaluation*. International Edition, 1992.
12. Khaled Khasawneh et al. SafeSpec: Banishing the Spectre of a Meltdown with leakage-free speculation. In *DAC*, page 60, 2019.
13. Khaled Khasawneh et al. SafeSpec: Banishing the Spectre of a Meltdown with leakage-free speculation. In *DAC*, page 60, 2019.
14. Vladimir Kiriansky et al. DAWG: A defense against cache timing attacks in speculative execution processors. In *MICRO*, pages 974–987, 2018.
15. Paul Kocher. "Spectre mitigations in Microsoft's C/C++ compiler. *MicrosoftCompilerSpectreMitigation. html*, 2018.
16. Paul Kocher, Jann Horn, et al. Spectre attacks: exploiting speculative execution. *Commun. ACM*, 63(7):93–101, 2020.
17. David Korczynski. Oss-fuzz: Continuous fuzzing for open source software. https://github.com/google/oss-fuzz-continuous-fuzzing-for-open-source-software, 2023.

18. Congmiao Li and Jean-Luc Gaudiot. Online detection of Spectre attacks using microarchitec-
    tural traces from performance counters. In *SBAC-PAD*, pages 25–28, 2018.
19. Congmiao Li and Jean-Luc Gaudiot. Challenges in detecting an "evasive spectre". *Com. Arch.
    Letters*, 19(1):18–21, 2020.
20. Peinan Li et al. Conditional speculation: An effective approach to safeguard out-of-order
    execution against Spectre attacks. In *HPCA*, 2019.
21. Moritz Lipp et al. Meltdown: reading kernel memory from user space. *Commun. ACM*,
    63(6):46–56, 2020.
22. Marc Löw. Overview of Meltdown and Spectre patches and their impacts. *Advanced
    Microkernel OS*, page 53, 2018.
23. Yangdi Lyu and Prabhat Mishra. A survey of side-channel attacks on caches and countermea-
    sures. *Journal of Hardware and Systems Security*, 2(1):33–50, 2018.
24. Yangdi Lyu and Prabhat Mishra. Scalable activation of rare triggers in hardware trojans
    by repeated maximal clique sampling. *IEEE Transactions on Computer-Aided Design of
    Integrated Circuits and Systems*, 40(7):1287–1300, 2020.
25. Linux Kernel Maintainers. Page table isolation, 2020.
26. Lars Müller. KPTI a mitigation method against meltdown. *Advanced Microkernel OS*, page 41,
    2018.
27. Zhixin Pan and Prabhat Mishra. Automated detection of Spectre and Meltdown attacks using
    explainable machine learning. In *2021 IEEE International Symposium on Hardware Oriented
    Security and Trust (HOST)*, pages 24–34. IEEE, 2021.
28. Zhixin Pan and Prabhat Mishra. Automated test generation for hardware trojan detection using
    reinforcement learning. In *Asia and South Pacific Design Automation Conference (ASPDAC)*,
    pages 408–413, 2021.
29. Zhixin Pan, Jennifer Sheldon, and Prabhat Mishra. Hardware-assisted malware detection
    and localization using explainable machine learning. *IEEE Transactions on Computers*,
    71(12):3308–3321, 2022.
30. Zhixin Pan, Jennifer Sheldon, Chamika Sudusinghe, Subodha Charles, and Prabhat Mishra.
    Hardware-assisted malware detection using machine learning. In *Design Automation and Test
    in Europe (DATE)*, 2021.
31. Marco Ribeiro et al. Why should I trust you?: Explaining the predictions of any classifier. In
    SIGKDD, pages 1135–1144, 2016.
32. Alvin E Roth. *The Shapley value: essays in honor of Lloyd S. Shapley*. Cambridge University
    Press, 1988.
33. Michael Schwarz, Moritz Lipp, Claudio Canella, Robert Schilling, Florian Kargl, and Daniel
    Gruss. Context: A generic approach for mitigating Spectre. In *NDSS*, 2020.
34. Guanhua Wang, Sudipta Chattopadhyay, Ivan Gotovchits, Tulika Mitra, and Abhik Roy-
    choudhury. oo7: Low-overhead defense against Spectre attacks via program analysis. *IEEE
    Transactions on Software Engineering*, 47(11):2504–2519, 2019.
35. Guanhua Wang et al. oo7: Low-overhead defense against Spectre attacks via binary analysis.
    CoRR, abs/1807.05843, 2018.
36. Hasini Witharana, Yangdi Lyu, and Prabhat Mishra. Directed test generation for activation of
    security assertions in RTL models. *ACM Transactions on Design Automation of Electronic
    Systems (TODAES)*, 26(4):1–28, 2021.
37. Mengjia Yan et al. InvisiSpec: Making speculative execution invisible in the cache hierarchy.
    In *MICRO*, page 1076, 2019.
38. Quanshi Zhang, Ruiming Cao, Ying Nian Wu, and Song-Chun Zhu. Growing interpretable
    part graphs on convnets via multi-shot learning. In *Proceedings of the AAAI Conference on
    Artificial Intelligence*, volume 31, 2017.
39. Quanshi Zhang, Yu Yang, Haotian Ma, and Ying Nian Wu. Interpreting CNNs via decision
    trees. In *Proceedings of the IEEE/CVF Conference on Computer Vision and Pattern
    Recognition*, pages 6261–6270, 2019.

Part III
Detection of Hardware Vulnerabilities

Hardware Trojan Detection Using Reinforcement Learning

1 Introduction

A vast majority of semiconductor companies rely on global supply chain to reduce design cost and meet time-to-market deadlines. The benefit of globalization comes with the cost of security concerns. For example, a typical automotive System-on-Chip (SoC) consists of about 100 Intellectual Property (IP) cores, and some of these cores may come from potentially untrusted third-party suppliers. An attacker may be able to introduce malicious implants in one of these third-party Ips. Hardware Trojan (HT) is a malicious modification of the target integrated circuit (IC) with two critical parts, trigger and payload. When the trigger is activated, the payload enables the malicious activity. For example, in Fig. 1, when the output of the trigger logic is true, the output of the payload XOR gate will invert the expected output. The trigger is typically created using a combination of rare events (such as rare signals or rare transitions) to stay hidden during normal execution. The payload represents the malicious impact that an HT will inflict to the target design, commonly resulting in information leakage or erroneous execution. Due to stealthy nature of these Trojans, it is infeasible to detect them using traditional functional validation methods. It is vital to detect HTs to enable trustworthy computing using modern SoCs.

There are many promising research efforts for malicious attack detection [14–18, 21]. These approaches can be broadly classified into two categories: side-channel analysis and simulation-based validation (logic testing). Side-channel analysis focuses on the difference in side-channel signatures (such as power, path delay, etc.) between the expected (golden specification) and actual (Trojan-inserted implementation) values [6, 10, 17]. A major drawback in side-channel analysis is that it is difficult to detect the negligible side-channel difference caused by a tiny Trojan (e.g., few gates in a multi-million gate design) since the difference can easily hide in process variation and environmental noise. In contrast, logic testing is

Z. Pan, P. Mishra, *Explainable AI for Cybersecurity*,
https://doi.org/10.1007/978-3-031-46479-9_5

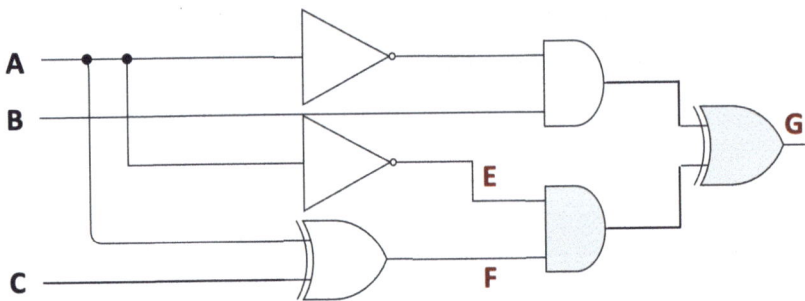

Fig. 1 An example hardware Trojan constructed by a trigger logic. When the output of the trigger logic (gray AND gate) is true, the output of the payload (gray XOR gate) will invert the expected output

robust against process variation and noise margins [3]. However, it is a fundamental challenge to activate an extremely rare trigger without trying all possible input sequences. Due to exponential input space complexity, traditional logic testing is not suitable for Trojan detection in large designs. Existing logic testing-based Trojan detection approaches have two fundamental limitations: high computation complexity (long test generation time) and low Trojan detection accuracy (low trigger coverage).

In this chapter, we describe an efficient logic testing approach for HT detection that addresses the following two challenges. (1) Existing logic testing approaches suffer from high computation complexity due to the fact that they require continuously flipping bits [4] of test vectors in an ad hoc manner to maximize the number of triggered rare activities. In contrast, we utilize a stochastic reinforcement learning framework to enable fast and automated generation of effective tests. (2) Existing approaches provide poor trigger coverage since they only focus on rare signals. This framework considers both rareness and the testability of signals using a combination of Sandia Controllability/Observability Analysis Program (SCOAP) measurement and dynamic simulation. It is expected to significantly improve the coverage of suspicious nodes with high stability.

2 Background and Related Work

In this section, we first survey related efforts on hardware Trojan detection using logic testing. Next, we provide a brief overview of reinforcement learning.

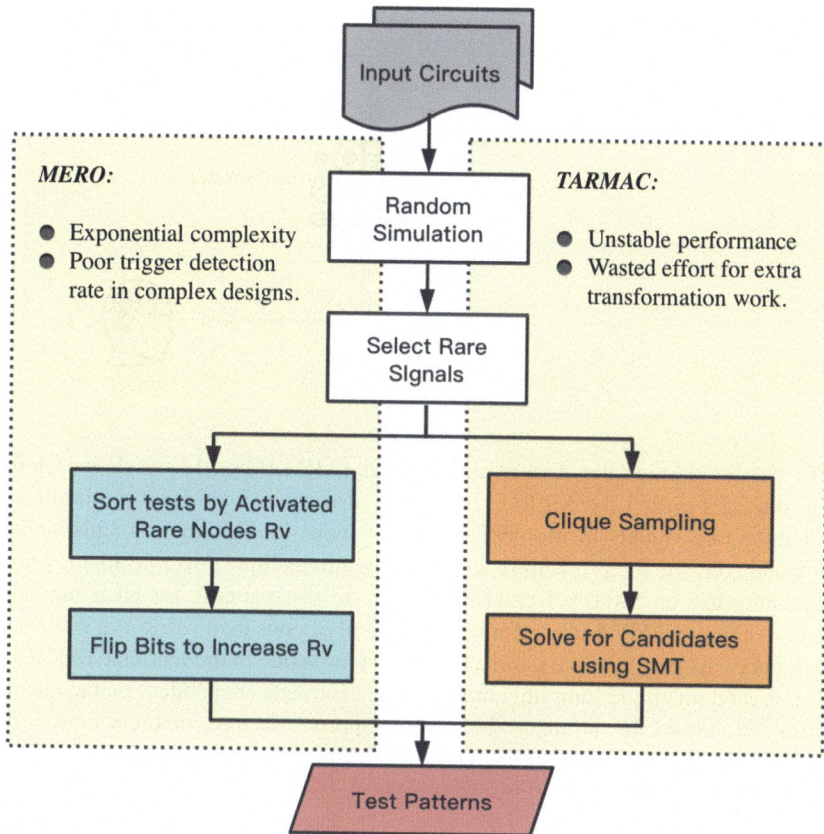

Fig. 2 Overview of state-of-the-art logic testing techniques: MERO [4] and TARMAC [9]

2.1 Logic Testing for Hardware Trojan Detection

The basic idea of logic testing for Trojan detection is to generate test patterns that are likely to activate the trigger conditions. In early days, random test generation was widely explored in industry due to its simplicity. However, there is no guarantee for activating stealthy Trojans using millions of random or constrained-random tests. MERO [4] proposed a statistical test generation scheme, which adopts the N-detect idea [19] to achieve better coverage. The heuristic behind is that if all rare signals are activated for at least N times, it is likely to activate the rare trigger conditions when N is sufficiently large. The left side of Fig. 2 shows an overview of MERO. It starts with random test generation followed by a brute-force process of flipping bits to increase the number of rare values being satisfied. It provides promising result for small benchmarks, but it introduces long execution time and scalability concerns, making it unsuitable for large benchmarks [9].

Fig. 3 Reinforcement
learning consists of five
important components: agent,
action, environment, state,
and reward

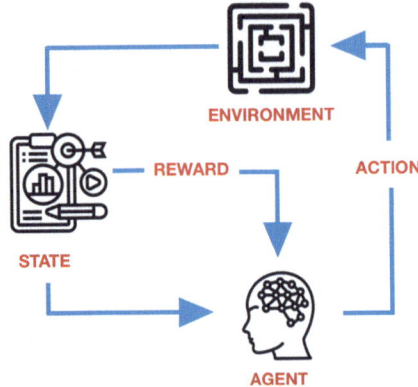

To address these issues, Lyu et al. proposed TARMAC [9, 11] as shown on the
right side of Fig. 2. Like MERO, TARMAC also starts with random simulation
to identify rare signals in the netlist. Next, it maps the design to a satisfiability
graph and converts the problem of satisfiability into a clique cover problem, where
the authors use an SMT solver [13] to generate test patterns for each maximal
clique. Although TARMAC performs significantly better than MERO in evaluated
benchmarks, its performance is very unstable. This is due to the fact that TARMAC
relies on random clique sampling, making its performance dependent on the quality
of sampled cliques. In summary, the existing approaches have inherent limitations
in terms of Trojan detection accuracy as well as test generation complexity.

2.2 Reinforcement Learning

Reinforcement learning [20] has earned its reputation as an efficient tool solving
problems with large and complex searching space [8]. Unlike traditional supervised
learning schemes, the training process of reinforcement learning is similar to the
nature of human learning. Basically, reinforcement learning works in an *adaptive*
way as shown in Fig. 3. There are five core components in reinforcement learning:
Agent, Action, Environment, State, and Reward. Reinforcement learning starts with
the interaction between agent and environment. At each step, agent utilizes its inner
strategy to decide the action to take, and the environment reacts to this action to
update the current state, which accordingly provides a reward value as feedback.
By giving positive reward for beneficial actions and penalty for inferior choices,
it allows the machine to distinguish the merits of certain action. Moreover, the
agent's strategy gets updated after receiving the feedback and tries to maximize
possible reward next time. Through continuous trials and rewards, the system
gradually adapts itself to make the most beneficial decisions, which quickly leads to
a desirable solution.

There are two key obstacles in directly applying this naive framework in test generation for Trojan detection.

1. *Reward Function:* Explicitly setting up proper reward for actions in test generation is difficult. For example, just counting the number of activated rare nodes is not a good metric to assign reward because an attacker may take multiple dimensions (such as rareness, controllability, observability, etc.) into account while designing a trigger condition.
2. *Action Space:* For a given n-bit test pattern, there are $2^n - 1$ possible ways to produce variations. It is impractical to meet both time and space requirements for dealing with such exponential action space.

The framework described in the next section addresses these challenges to generate test patterns for detecting hardware Trojans.

3 Test Generation Using Reinforcement Learning

Figure 4 shows an overview of our proposed test generation scheme using reinforcement learning (TGRL). For a given circuit design, we first apply a combination of static testability analysis and dynamic simulation, where the simulation provides us with information of rare nodes (Sect. 3.1) and the testability analysis computes SCOAP testability parameters (Sect. 3.2) of each node in the circuit. Next, these intermediate results are fed into the machine learning model as primary inputs. We utilize reinforcement learning (RL) as the learning model due to its outstanding potential in efficiently solving problems with large and complex solution space [8]. The reinforcement learning model is trained with a stochastic learning scheme to generate test vectors, and it continuously improves itself to cover as many suspicious nodes as possible. After sufficient iterations of training, the trained RL model is utilized for automatic test generation (Sect. 3.3). It starts with initial input patterns and continuously generates a set of test patterns until we get the required number of test patterns.

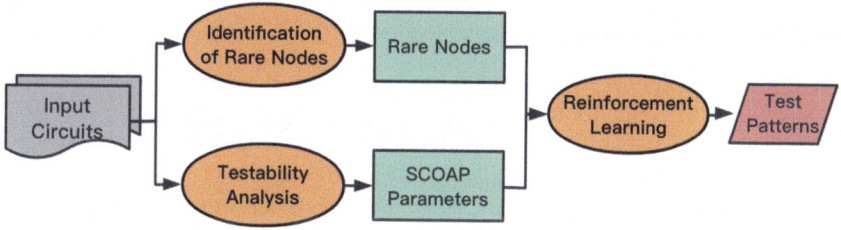

Fig. 4 Overview of our test generation framework that consists of three major activities: identification of rare nodes, testability analysis, and reinforcement learning

3.1   Identification of Rare Nodes

Like existing approaches, we utilize the dynamic simulation of the benchmark to identify rare nodes. First, the design needs to be simulated using a reasonable number of random or constrained-random test patterns. Next, the trace needs to be analyzed to determine how many times each node (signal) is assigned a value "0" or "1" during the simulation. Finally, we need to select the signals (with specific values) as rare nodes that are below a specific threshold. For example, if the output of the NOR gate was "0" 96% of the time (i.e., "1" with 4% of the time) during simulation and threshold is 5%, the output of the NOR gate with value "1" will be marked as a rare node. A threshold is considered reasonable if the trigger constructed by the respective rare nodes cannot be covered by traditional simulation-based validation using millions of random tests.

3.2   Testability Analysis

While majority of existing techniques mainly consider rareness to evaluate suspicious signals, it remains the responsibility of the defender to come up with a more comprehensive measurement. In our approach, we exploit Sandia Controllability/Observability Analysis Program (SCOAP), which takes both *controllability* and *observability* attributes of signals into consideration. In essence, controllability indicates the amount of effort required for setting a signal to a specific value, while observability weighs up the difficulty of propagating the target signal toward observation points.

The testability measurement naturally fits the demand of HT detection from a security perspective. Clearly, signals with low controllability are more likely to be chosen as trigger signals. Because low controllability guarantees the difficulty of switching these signals with a limited number of test patterns. Similarly, targeting signals with low observability as payload are favorable for attackers, since it coincides with HT's clandestine property and avoids them from frequently generating observable impact on design outputs.

The SCOAP method quantifies the controllability and observability of each signal in the circuit with three numerical values.

- **CC0**: Combinational 0-controllability, the number of signals must be manipulated to set "0" value for target.
- **CC1**: Combinational 1-controllability, the number of signals must be manipulated to set "1" value for target.
- **CO**: Combinational observability, the number of signals must be manipulated to observe target value at primary outputs.

The SCOAP computation can be performed in a recursive manner. First, the boundary conditions are the primary inputs (PIs) and primary outputs (POs), where

Logic gates	Testability Measure
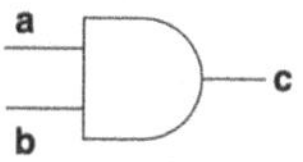 a b ─ c (AND gate)	$CC0(c) = min\{CC0(a), CC0(b)\} + 1$ $CC1(c) = CC1(a) + CC1(b) + 1$ $CO(a) = CO(c) + CC1(b) + 1$
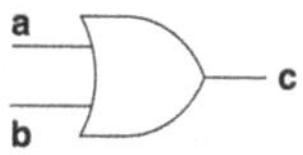 a b ─ c (OR gate)	$CC0(c) = CC0(a) + CC0(b) + 1$ $CC1(c) = min\{CC1(a), CC1(b)\} + 1$ $CO(a) = CO(c) + CC0(b) + 1$
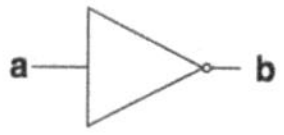 a ─ b (NOT gate)	$CC0(a) = CC1(b) + 1$ $CC1(a) = CC0(b) + 1$ $CO(a) = CO(b) + 1$

Fig. 5 Formula of SCOAP testability measurement for three fundamental logic gates

$$CC0(PI) = CC1(PI) = 1$$

$$CO(PO) = 0$$

This is straightforward, since only one manipulation is required for controlling primary input (itself), while no extra operation is needed for observing primary output. Next, the circuit is converted into a directed acyclic graph (DAG) and further levelized by topological sorting. For each gate, the output controllability is determined by controllability of its inputs, while the input observability is determined by observability of output and all the other input signals. Figure 5 shows the computation formula for three fundamental logic gates. Consider the $CC1$ measurement of AND gate as an example, and in order to control the output signal c as "1," both of its input signals a and b should be manipulated as "1" at the same time. Therefore, we have $CC1(c) = CC1(a) + CC1(b) + 1$, where the "+1" is for counting the level depth. It is worth noting that since the controllability parameters of inputs are necessary for computing that of the output signal, the SCOAP procedure starts from calculating controllability values for all signals in a direction from PI toward PO. Afterward, signals' observability is measured in the reverse direction.

The task of SCOAP testability analysis can be performed in parallel with identification of rare signals in the circuit. The computed attributes (SCOAP parameters and rare signal values) will be fed into a reinforcement learning model to fulfill automatic test generation as discussed in the next section.

3.3 Utilization of Reinforcement Learning

Based on the workflow and challenges for reinforcement learning, we present our learning paradigm by listing the mapping from objects in the test generation task onto the five crucial components of reinforcement learning.

Agent Agent refers to the object interacting with the environment. In our test generation problem, it is chosen as the current test vector under processing and we denote it as t.

Environment Circuit design is mapped into environment, which receives the input test vector to produce meaningful results. We denote it as D.

State State refers to information presented by the environment that can be perceived by the user, such as conditions and parameters. We map the SCOAP parameters and rare signal values of the entire circuit as state. They are encoded by two functions rv and $scoap$, where rv returns the rare value for a specific signal, and $scoap$ is defined as follows:

$$scoap(s) = | < CO(s), CC(rv(s)) > |$$

For a given signal s, $CO(s)$ is the combinational observability of s, and $CC(rv(s))$ is the combinational controllability corresponding to the rare value of s. We are utilizing the $L-1$ norm of SCOAP parameters to measure the synthesized testability of signal s. The state records the basic information of the interaction between the current test vector and the circuit, which can be further utilized in reward computation.

Action Action space consists of all possible operations that make changes to the system. For the test generation problem, a natural choice is the total set of possible bit-flipping operations. However, in that case the action space (size) for a vector of length n is $2^n - 1$, which is impractical for encoding and manipulation. We apply a stochastic approach to address this challenge as described in Fig. 6. In our approach, for each bit in the current test vector, a probabilistic selection will determine whether to flip it or not. In other words, the action is chosen randomly at each step. This non-deterministic action is not completely arbitrary but determined by the given probability distributions, which guarantees the coverage of all possible flipping operations.

This approach sheds light on drastically reducing the cost for encoding actions. The probabilistic selection happened on each bit is a binary selection, which can be encoded by one floating-point number. Therefore, an n-bit test pattern requires a vector function $P(\theta) = [\theta_1, \theta_2, \ldots, \theta_n]$ to formulate the entire space of probability selection.

Reward In general, reward value is the most important feedback information from the environment that describes the effect of the latest action. For optimization problems, it often refers to the benefit of performing the current operation. In our framework, we exploit a composite reward evaluation scheme consisting of two components, **_rare reward_** and **_testability reward_**.

Given the current test vector t and action space P, we first denote the newly generated test vector as $t_p = act(t, P)$, and $D(s, t_p)$ as the value of signal s after applying t_p for D, by which we can further define the reward value R in test generation as follows:

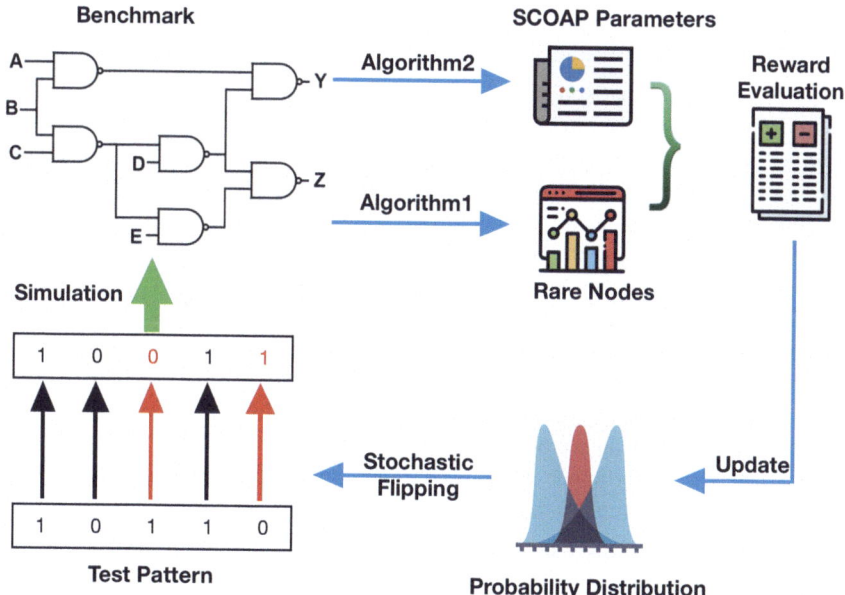

Fig. 6 Overview of stochastic reinforcement learning for test generation

$$R_r(t_p) = size(\{s|D(s, t_p) = rv(s)\})$$

$$R_t(t_p) = \sum scoap(s) \quad w.r.t \ D(s, t_p) = rv(s)$$

$$R(t_p) = R_r(t_p) + \lambda \cdot R_t(t_p)$$

Here, $R_r(t_p)$ is the rare reward, which is based on the counting of the number of triggered rare signals, and $R_t(t_p)$ is the testability reward defined as the summation of $scoap$ measurement of corresponding signals. Finally, we put $\lambda \in \mathbb{R}^+$ as a regularization factor to balance the weight of two components. This reward value is exploited in a reinforcement learning model to update hyperparameters at each iteration representing interaction between "agent" and "environment." Specifically, we apply propagation [7] with the computed reward value to adjust those probability distributions, and when a positive reward is obtained, the probability of the corresponding action is increased and vice versa.

4 Experiments

In this section, we first outline the experimental setup. Next, we present the experimental results in terms of trigger coverage as well as test generation time.

4.1 Experimental Setup

To enable a fair comparison with existing approaches, we deploy the experiment on the same benchmarks as [4, 9] from ISCAS-85 and ISCAS-89 [1]. Also, we preserve the parameter configuration applied in those papers, where the rareness threshold is set to 0.1, and the total number of sampled Trojans is 1000. The code for benchmark parsing and identification of rare nodes is written in C++17. To perform SCOAP analyses, we use the open-source Testability Measurement Tool from [2]. The reinforcement learning model in our approach was conducted on a host machine with Intel i7 3.70 GHz CPU, 32 GB RAM, and RTX 2080 256-bit GPU. We choose Python (3.6.7) code using PyTorch (1.2.0) with https://www.overleaf.com/project/5f179de1547e3b0001d694d2 cudatoolkit (10.0) to implement the machine learning framework. The training process consisted of 500 epochs, where we initialize the learning rate α as 0.02 at the beginning and lower it down to 0.01 after 200 epochs. We compare the performance in terms of trigger coverage and test generation time between the following methods:

- **MERO:** statistical test generation for Trojan detection utilizing multiple excitation of rare occurrences [4]
- **TARMAC:** the state-of-the-art test generation method for Trojan detection using clique cover [9]
- **TGRL:** our proposed test generation technique for Trojan Detection using reinforcement learning

4.2 Results on Trigger Coverage

Table 1 demonstrates the effectiveness of our proposed methods compared to the state-of-the-art methods. The first column lists the benchmarks. The second column shows the number of signals in those designs. The third, fifth, and seventh columns provide the number of tests generated by MERO [4], TARMAC [9], and our approach, respectively. The fourth, sixth, and eighth columns show the trigger coverage using the tests generated by MERO [4], TARMAC [9], and our approach, respectively. The last two columns present the improvement in trigger coverage provided by our approach compared to the state-of-the-art methods. Clearly, MERO provides a decent trigger coverage on tiny designs such as c6288, while its trigger coverage drastically drops to less than 10% when applied to large designs like s15850. TARMAC provides promising improvement compared with MERO, but we can observe that it does not have a consistent outcome. For example, in case of c6288 and c7552 with comparable size, the trigger coverage drastically differs (86.1% versus 58.7%). Such huge gap clearly indicates TARMAC's instability with respect to various benchmarks. In contrast, our approach achieves 100% trigger coverage for the first three benchmarks. When we consider large designs, our approach still maintains a high trigger coverage. Overall, our approach outperforms both MERO

Table 1 Comparison of trigger coverage with existing approaches

Benchmarks	MERO [4]		TARMAC [9]		Proposed approach (TGRL)	
	# Tests	Trigger-Cov(%)	# Tests	Trigger-Cov(%)	# Tests	Trigger-Cov(%)
c2670	6820	33.1	6820	100	6820	100
c5315	9232	54.3	9232	84.6	9232	100
c6288	5044	68.9	5044	86.1	5044	100
c7552	14914	4.9	14914	58.7	14914	97.3
s13207	44534	2.6	44534	84.2	44534	93.4
s15850	39101	2.2	39101	66.3	39101	88.5
s35932	34041	8.6	34041	91.5	34041	93.7
Average	21955	24.99	21955	81.62	21955	96.12

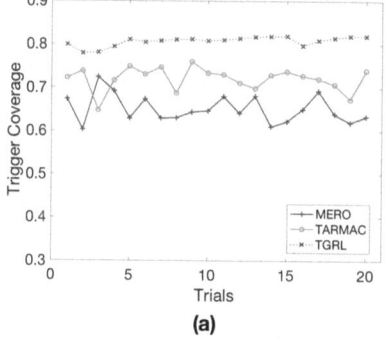 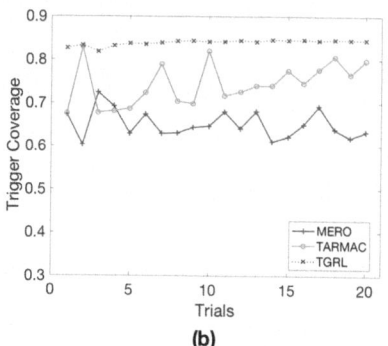

(a)   (b)

Fig. 7 The variation of trigger coverage in 20 trials. TARMAC and MERO demonstrate unstable performance, while our approach provides consistently high trigger coverage. (**a**) c7552. (**b**) s15850

(up to 92.4% and 77.1% on average) and TARMAC (up to 38.6% and 14.5% on average) in trigger coverage.

Table 1 also reveals the weakness of previous works in terms of "stability" in trigger coverage. To confirm our observation, we further evaluate the stability of all approaches. We choose c7552 and s15850 as target benchmarks, where we repeat each approach for 20 trials and record the trigger coverage, in order to study the extent of variations. The results are shown in Fig. 7. As we can see from the figure, our proposed method preserves stable performance across 20 trials. However, there are drastic variations in trigger coverage for the other two approaches. Especially when applied to larger benchmark like s15850, this phenomenon becomes more obvious. The standard deviation of TARMAC is high (0.1876), while it is negligible for our proposed method (0.0237). In reality, a stable performance is desirable; otherwise a user needs to try numerous times to obtain an acceptable result, which can be infeasible for Trojan detection in large designs.

Table 2 Comparison of Test Generation Time (in seconds)

Design	MERO	TARMAC	TGRL	MERO/TGRL	TARMAC/TGRL
c2670	1149	301	74	15.52x	4.06x
c5315	3791	643	126	30.08x	5.11x
c6288	826	666	108	7.64x	6.16x
c7552	7423	2809	169	43.92s	16.62x
s13207	16508	6022	1328	12.4x	4.53x
s15850	16429	12580	1204	13.64x	10.44x
s35932	53171	23446	4092	12.99x	5.72x
Average	14185	6638	1014	14.1x	6.54x

4.3   Results on Test Generation Time

Table 2 compares the test generation time for the three methods. The first column lists the benchmarks. The next three columns provide the test generation time for MERO [4], TARMAC [9], and our approach, respectively. The last two columns show the time improvement provided by our approach compared to the other methods.

Clearly, our approach provides the best results across benchmarks, while MERO is the worst. Not surprisingly, MERO lags far behind the other two in time efficiency due to its brute-force bit-flipping method. While TARMAC provides better test generation time than MERO, our approach is significantly (6.54x on average) faster than TARMAC. There are three major bottlenecks that slow down TARMAC. First, TARMAC requires extra transformation to map the circuit design into a satisfiability graph. Next, the clique sampling in TARMAC is compute-intensive, and it repeatedly removes nodes from circuit and re-computes logic expression for each potential trigger signal. Finally, TARMAC exploits an SMT solver to generate each candidate test vector, which determines the upper bound of its time efficiency. In contrast, our proposed approach does not use any satisfiability solver. Only overhead in our approach is the training time—the model training is composed of 500 iterations where each iteration is basically a one-step test mutation and evaluation. When the model is well trained, it can automatically generate all the remaining test vectors without extra efforts. Note that the reported test generation time includes the model training time. Overall, our proposed approach drastically (up to 16.6x, 6.54x on average) improves the test generation time compared to state-of-the-art methods.

5   Summary

The detection of hardware Trojans is an emerging and urgent need to address semiconductor supply chain vulnerabilities. While there are promising test generation techniques [5, 12, 22], they are not useful in practice due to their inherent fundamental limitations. Specifically, they cannot provide reasonable trigger coverage. Most importantly, they require long test generation time and still provide unstable performance. To address these serious challenges, this chapter described an automated test generation scheme using reinforcement learning for effective hardware Trojan detection. Specifically, it explored an efficient combination of rareness of signals and testability attributes to provide a fresh perspective on improving the coverage of suspicious signals. It also developed an automated test generation scheme utilizing reinforcement learning model trained with stochastic methods which is able to drastically reduce the test generation time. Experimental results demonstrated that our approach can drastically reduce the test generation time (6.54x on average), while it is able to detect a vast majority of the Trojans in all benchmarks (96% on average), which is a significant improvement (14.5% on average) compared to state-of-the-art methods.

References

1. ISCAS benchmarks. https://filebox.ece.vt.edu/~mhsiao/iscas89.html.
2. Scoap. https://sourceforge.net/projects/testabilitymeasurementtool/.
3. Alif Ahmed, Farimah Farahmandi, Yousef Iskander, and Prabhat Mishra. Scalable hardware trojan activation by interleaving concrete simulation and symbolic execution. In *International Test Conference*, pages 1–10, 2018.
4. Rajat Subhra Chakraborty, Francis Wolff, Somnath Paul, Christos Papachristou, and Swarup Bhunia. MERO: A statistical approach for hardware trojan detection. In *CHES*, pages 396–410, 2009.
5. Farimah Farahmandi and Prabhat Mishra. Automated test generation for debugging multiple bugs in arithmetic circuits. *IEEE Transactions on Computers*, 68(2):182–197, 2018.
6. Yuanwen Huang, Swarup Bhunia, and Prabhat Mishra. Scalable test generation for trojan detection using side channel analysis. *IEEE Transactions on Information Forensics and Security (TIFS)*, 13(11):2746–2760, 2018.
7. Henry J Kelley. Gradient theory of optimal flight paths. *Ars Journal*, 30(10):947–954, 1960.
8. Sami Khairy, Ruslan Shaydulin, Lukasz Cincio, Yuri Alexeev, and Prasanna Balaprakash. Reinforcement-learning-based variational quantum circuits optimization for combinatorial problems. *CoRR*, abs/1911.04574, 2019.
9. Yangdi Lyu and Prabhat Mishra. Automated trigger activation by repeated maximal clique sampling. In *ASPDAC*, pages 482–487, 2020.
10. Yangdi Lyu and Prabhat Mishra. Maxsense: Side-channel sensitivity maximization for trojan detection using statistical test patterns. *TODAES*, 2020.
11. Yangdi Lyu and Prabhat Mishra. Scalable activation of rare triggers in hardware trojans by repeated maximal clique sampling. *IEEE Transactions on Computer-Aided Design of Integrated Circuits and Systems*, 40(7):1287–1300, 2020.
12. Yangdi Lyu, Xiaoke Qin, Mingsong Chen, and Prabhat Mishra. Directed test generation for validation of cache coherence protocols. *IEEE Transactions on Computer-Aided Design of Integrated Circuits and Systems*, 38(1):163–176, 2018.
13. L. Moura and N. Bjørner. Z3: an efficient SMT solver. In *TACAS*, pages 337–340, 2008.
14. Zhixin Pan and Prabhat Mishra. Automated detection of spectre and meltdown attacks using explainable machine learning. In *2021 IEEE International Symposium on Hardware Oriented Security and Trust (HOST)*, pages 24–34. IEEE, 2021.

15. Zhixin Pan and Prabhat Mishra. A survey on hardware vulnerability analysis using machine learning. *IEEE Access*, 10:49508–49527, 2022.

16. Zhixin Pan and Prabhat Mishra. Ai trojan attack for evading machine learning-based detection of hardware trojans. *IEEE Transactions on Computers*, 2023.

17. Zhixin Pan, Jennifer Sheldon, and Prabhat Mishra. Test generation using reinforcement learning for delay-based side-channel analysis. In *IEEE/ACM International Conference On Computer Aided Design (ICCAD)*, pages 1–7, 2020.

18. Zhixin Pan, Jennifer Sheldon, and Prabhat Mishra. Hardware-assisted malware detection and localization using explainable machine learning. *IEEE Transactions on Computers*, 71(12):3308–3321, 2022.

19. Irith Pomeranz and Sudhakar M. Reddy. A measure of quality for n-detection test sets. *IEEE Trans. Computers*, 53(11):1497–1503, 2004.

20. Stuart J Russell and Peter Norvig. *Artificial intelligence: a modern approach*. Malaysia; Pearson Education Limited,, 2016.

21. Hasini Witharana, Aruna Jayasena, Andrew Whigham, and Prabhat Mishra. Automated generation of security assertions for RTL models. *ACM Journal on Emerging Technologies in Computing Systems*, 19(1):1–27, 2023.

22. Zhixin Pan and Prabhat Mishra. Hardware trojan detection using Shapley ensemble boosting. pages 1127–1130, 2021.

Hardware Trojan Detection Using Side-Channel Analysis

1 Introduction

With the rapid development of semiconductor technologies coupled with increasing demands of complex *System-on-Chips* (SoCs), the vast majority of semiconductor companies utilize global supply chains. A long and distributed supply chain provides opportunity for third-party Intellectual Property (IP) vendors as well as service providers to implant hardware Trojans (HTs) inside SoCs [7, 22, 24, 25]. Therefore, Trojan detection is widely acknowledged as a major focus to enable secure and trustworthy SoCs.

Existing Trojan detection techniques can be broadly classified into two categories: logic testing and side-channel analysis. Logic testing methods such as Automatic Test Pattern Generation (ATPG) [2] or statistical test generation [4, 14] try to activate Trojans using generated tests, but they have two major limitations: (1) They suffer from high computational complexity for large designs. (2) Since it is infeasible to generate all possible input patterns, the generated tests are not effective in activating stealthy Trojans. The triggering conditions for Trojans are usually crafted as a combination of rare conditions, such that Trojan-implanted designs will retain exactly the same functionality as golden designs until a rare condition is satisfied to yield malicious behavior. Figure 1 shows a simple example from ISCAS'85 benchmark. In this design, both F and G are signals with "0" as the rare value. The shaded AND gate with inverted inputs will only be triggered if both F and G become "0," and, once triggered, the succeeding XOR gate (payload) will invert signal I, which leads to change in functionality.

The side-channel analysis is a promising alternative since it compares the difference of side-channel signatures (such as path delay, electromagnetic emanation, dynamic current, etc.) with the expected (golden) values to detect Trojans. However, the effectiveness of side-channel analysis depends on the HT's side-channel leakage. The noise induced by the environment or process variation usually overshadows the Trojan footprint, which makes the detected difference negligible. Recent efforts

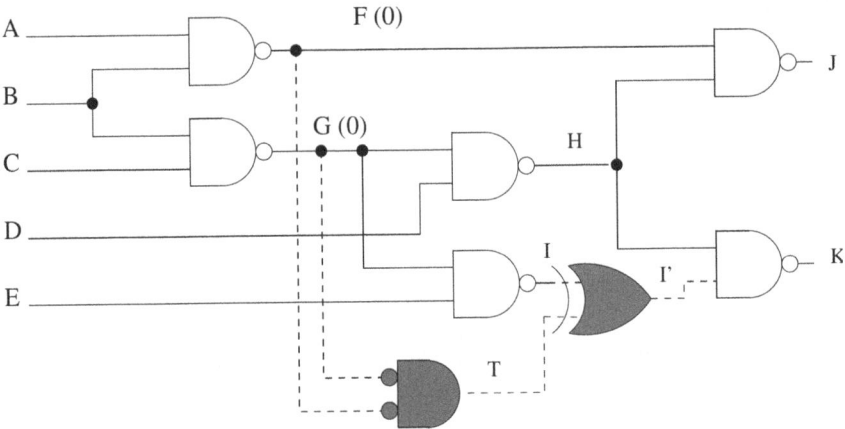

Fig. 1 An example of a hardware Trojan triggered by rare signals

have tried to combine logic testing and side-channel analysis [10, 11, 15, 17, 18] in order to improve side-channel sensitivity. Specifically, their goal is to maximize activity in suspicious regions while minimizing the activity in the rest of the design. While existing approaches provide promising avenues, they face two major challenges. First, the test generation time complexity is exponential for these methods, which severely limits their usability. Second, the side-channel difference achieved by these approaches is not large enough to offer high confidence in HT detection results.

In this chapter, we describe a test generation approach using reinforcement learning for delay-based side-channel analysis. Crafting test vectors that can reveal the impact of implanted Trojans on path delay is not a trivial task. The computational complexity grows exponentially with the design size. It is time consuming and even impractical for large designs due to the exponential nature of possible paths. The difficulty of delay-based side-channel HT detection also comes from designing proper test vectors to increase the observability of side-channel differences. Specifically, the test vectors should be able to reveal the impact of an inserted Trojan on path delay as much as possible. Existing approaches have focused on passively enumerating possible paths affected by the HT. But if the HT is not triggered, only a few gates difference can be obtained, which is hard to distinguish from environmental noise. Consequently, the detection results are not promising. In contrast, our proposed solution focuses on exploring the impacts through the critical path analysis, which allows us to actively change the critical path and magnify side-channel differences.

The remarkable success of machine learning (ML) demonstrates its flexibility and state-of-the-art performance in various domains [19–21, 28], and there are many ML application hardware security tasks [5, 6, 9], which inspired us to explore its potential in HT detection. Specifically, we describe an automated and efficient test generation method using reinforcement learning to maximize the difference in path delay between the Trojan-implanted design and the golden design by exploiting the critical path analysis. The experimental evaluation demonstrates that such a framework can outperform the state-of-the-art technique in both test generation time efficiency (17x on average) and side-channel sensitivity (59% on average).

2 Background and Motivation

In this section, we first provide an overview of reinforcement learning. Next, we briefly outline the motivation behind using delay-based side-channel analysis.

2.1 Background: Reinforcement Learning

Reinforcement learning (RL) has shown its potential in solving complex optimization problems [12, 16, 27]. Searching for optimal test vectors in target designs to maximize the side-channel sensitivity can be viewed as an optimization problem. RL is a branch of machine learning, but unlike the commonly known supervised learning, it is closer to human learning. For example, although an infant baby is not able to understand spoken words, it can still master language after a period of exploration. This exploration process is actually a process of gradually learning the rules (lexicon and grammar) of speaking through trials and responding to feedback from the environment. Similarly, RL also learns to find an optimal strategy through a series of attempts and constantly adjusts its behavior based on the feedback.

An overview of RL framework is shown in Fig. 2. It consists of five core components: *Agent, Environment, Action, State,* and *Reward*.

- **Agent** refers to the object that can interact with the environment through actions. The agent of reinforcement learning is usually the set of test cases to be optimized, which is continuously updated through learning process.

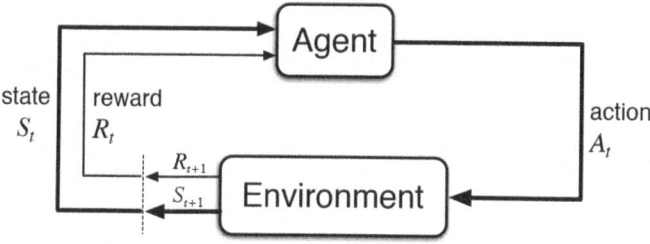

Fig. 2 The basic framework of reinforcement learning. At time stamp t, state S_t and reward R_t are fed into agent, which produce action A_t with updated strategy. The agent interacts with the environment to obtain new state S_{t+1} and reward R_{t+1} and then starts the next round of learning process

- *Environment* is the receiver of the action, such as the optimization problem itself.
- *Action* consists of all possible operations that may affect the environment, such as using the current strategy for one-step calculation.
- *State* refers to information about the environment that can be perceived by the agent, such as conditions and parameters.
- *Reward* is the feedback information from the environment that describes the effect of the latest action. For optimization problems, it often refers to the gain of objective function after performing the current operation.

The process of reinforcement learning is a process of obtaining feedback by interacting with the environment and then adjusting the actions based on the feedback in order to maximize the total reward. The goal of reinforcement learning is to find an optimal strategy to maximize the rewards obtained during the entire interaction process. In terms of implementation, reinforcement learning is a process of gradually optimizing the parameters of the algorithm through multiple rounds to enhance the learning effect. To the best of our knowledge, our work is the first attempt in utilizing reinforcement learning for side-channel-analysis-aware test generation.

2.2 Motivation: Delay-Based Side-Channel Analysis

There are various physical signatures of electronic devices suitable for side-channel analysis, such as path delay, dynamic current [18], and electromagnetic emanations [13]. Compared to the other side-channel signatures, path delay provides the following advantages:

1. *Independence:* The delay between any gates in the design can be measured independently, which provides more comprehensive information compared to other side-channel signals.
2. *Diversity:* Implanted Trojans can impact path delay in multiple ways. Let us consider the design in Fig. 1 as an example. There will be an increase of propagation delay for the gates producing signals F and G, since they are connected to an extra gate, which leads to increased capacitive load. Second, since one XOR gate and one AND gate were inserted to deliver the payload, the path delay will always have at least two gates difference from the golden design for any path through these inserted XOR/AND gates.
3. *Stability:* Delay-based Trojan detection techniques provide superior performance under parameter variations by leveraging statistical techniques [23]. This stability guarantees high confidence of detection results from delay-based analysis.
4. The detection result is extremely sensitive to environmental noise since the differences in delay induced by these methods are often negligible. Without activation of the HT, the difference is usually from a few gates. In fact, even with Trojan successfully triggered, they are not guaranteed to generate a critical path from the Trojan to an observable output for propagating the delay.

3   Reinforcement Learning-Based Path Delay Analysis

Figure 3 shows an example of maximizing delay difference by changing the critical path. The activation of the trigger T is necessary for maximizing the delay difference. The top part of the figure shows that if the test vector fails to activate the Trojan ($T = 0$), the critical path from the input layer toward the Trojan is exactly the same as that in the golden design. Then the delay difference is limited to the inserted gates themselves (e.g., only one XOR gate in the figure). The bottom part of the figure indicates that the critical path will be drastically different if the trigger can be activated. Note that the trigger signal T has to switch between consecutive input patterns; otherwise there will be no contribution from the Trojan to the path delay because the related signals remain the same between two consecutive tests.

Once the above requirements are satisfied, a completely different critical path from input layer to Trojan is obtained, so that we can expect a huge difference between the measured delay differences. Consequently, the big problem of test generation now is divided into two sub-problems: how to find a good initial test for triggering the Trojan and how to efficiently generate proper succeeding tests to switch triggering signals. Due to their stealthy nature, HTs are very likely to be activated by rare, triggered conditions, and therefore, the two sub-problems can be transformed into:

- Generate initial tests for triggering rare nodes
- Generate succeeding tests for triggering rare switches

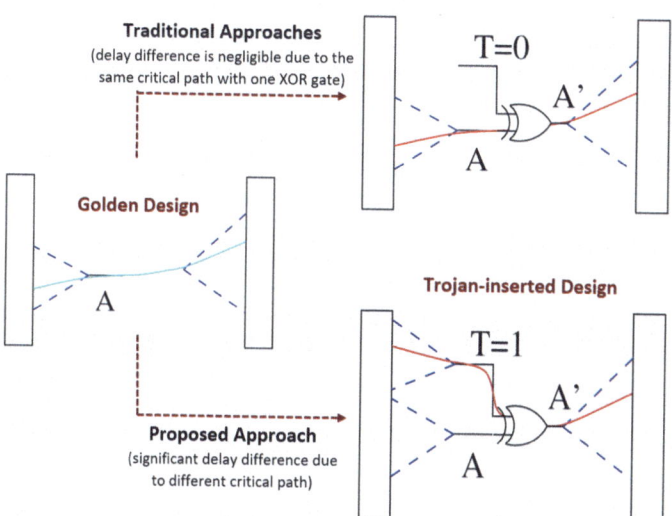

Fig. 3 Maximizing delay difference by changing the critical path. By triggering T, the critical path from input layer to A' is significantly changed

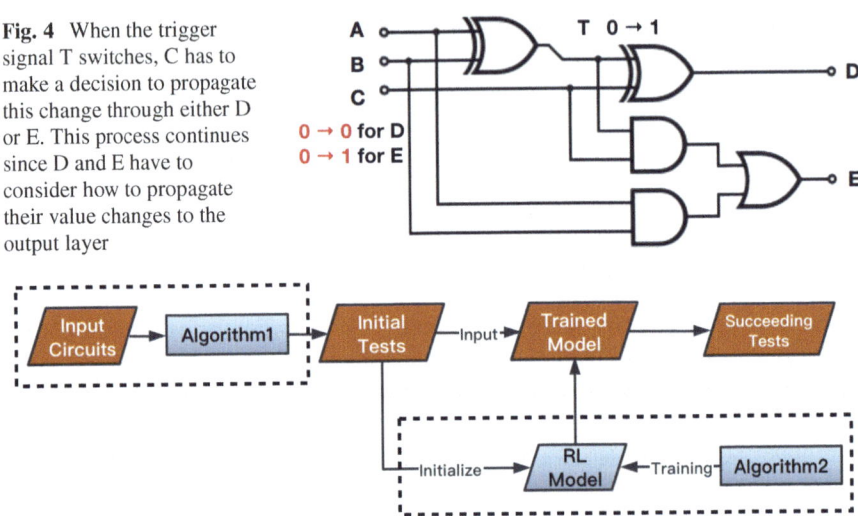

Fig. 4 When the trigger signal T switches, C has to make a decision to propagate this change through either D or E. This process continues since D and E have to consider how to propagate their value changes to the output layer

Fig. 5 Delay-based Trojan detection framework consists of two major activities: initial test generation and reinforcement learning

There are major research challenges in solving the above two sub-problems. If we shift focus to a succeeding path (a path from the Trojan to the output layer in which the delay is propagated through the design) starting from node, there has to be a critical path from it to the output layer to propagate the delay. Otherwise the delay is cut off and hidden from all other nodes succeeding ones. Unfortunately, creating such path is an NP-hard problem because, in the worst-case scenario, every signal in the critical path has to be taken into consideration. For example, in Fig. 4, the trigger signal T switches from 0 to 1. Signal C, whose original value is 0, can either switch to 1 for propagating the delay induced by the switch through the AND gate to E or remain unchanged to propagate it through the XOR gate to D. This process will recursively continue for D and E to calculate constraints if they are chosen to be added to the critical path. The search space grows exponentially, bringing in numerous constraints. Therefore, traditional SAT-based approach is not feasible here, and we plan to apply reinforcement learning to tackle this issue. We will discuss this application in detail in Sect. 3.3.

3.1 Overview

Figure 5 shows an overview of reinforcement learning-based side-channel analysis for the detection of hardware Trojans. The primary goal is to generate a sequence of test patterns $(t_1, t_2, \ldots, t_n)$ such that for every consecutive pair of tests (t_i, t_{i+1}), the delay-based side-channel sensitivity is maximized. For a given circuit design, we first obtain a set of proper initial test cases to ensure triggering rare nodes (Sect. 3.2). Next, those initial test cases in the previous step are fed into a reinforcement machine learning model as initial inputs, which is trained with a stochastic learning scheme (Sect. 3.3) to increase the probability of triggering rare switches. After sufficient iteration of training, a well-trained RL model is exploited for automatic test generation. It starts working with initial input patterns and utilizes the newly generated test vectors as input in the next round to continuously generate a sequence of test patterns of the desired amount.

3.2   Generation of Initial Vectors

As discussed before, it is important for test patterns to activate trigger conditions because if the test pattern fails to activate the Trojan, the delay of the golden design and the Trojan-inserted design differs by, at most, one gate. Therefore, our goal is to maximize the probability of activating trigger conditions. Since an attacker is likely to construct trigger conditions based on rare nodes to avoid detection, we need to generate initial vectors that can maximize the activation of rare nodes. It accepts the design (circuit netlist), the list of rare nodes (R), and the number of test vectors (n) as inputs and produces n test vectors. The number of vectors presents a trade-off—a larger number can lead to longer runtime, but it is likely to improve the probability of activating trigger conditions compared to a smaller number of initial vectors.

This algorithm first computes the logic equations for each rare node by analyzing the cone of influence for each of them. Next, it generates n test vectors, one in each iteration. The test generated in each iteration is expected to activate different rare nodes since we randomize the order of the rare nodes. This algorithm also tries to a generate a test vector that can cover a large number of rare nodes. We ensure this by adding as many rare nodes as possible without making an invalid trigger. We use a SAT solver [8] to find a test that would activate all the nodes in the trigger simultaneously. Since an attacker is likely to create a trigger with the smallest possible number of rare nodes to minimize the path delay footprint, tests generated by our approach have a higher likelihood of detecting small triggers. In other words, if we can find a test that can cover a trigger with a large number of nodes, although we do not know the actual trigger inserted by an attacker, the likelihood of activating that Trojan goes up if we assume that the actual trigger will consist of a small number of nodes because the small trigger signals are a subset of the large trigger signals. These test cases will be used by the reinforcement learning (RL) model to form the initial state. The RL model is able to automatically search for the best succeeding test patterns, as discussed in the next section.

Algorithm 1: Generation of initial test patterns

    Input   : Design (D), Rare nodes (R), Number of initial vectors (n)
    Output: Test Patterns
1 Compute logic equations for each rare node in D
2 Initialize $Tests=\{\}$
3 $i = 1$
4 **repeat**
5   Trigger $TR = \emptyset$
6   Randomize the order of rare nodes R
7   **for** *each rare node $r \in R$* **do**
8   **if** $TR \cup r$ *is a valid trigger* **then**
9 $TR = TR \cup r$

10   Solve TR and get a $test_{TR}$
11 $Tests = Tests \cup test_{TR}$
12 $i = i + 1$
13 **until** $i > n$;
14 Return $Tests$

3.3 Generation of Succeeding Vectors

There are two crucial requirements for a succeeding test vector (a test vector produced by the RL model, which can then be used to generate another succeeding, or consecutive, test vector). First, the rare signals triggering Trojans have to switch between consecutive test patterns. If there is no rare switch, the critical path will not pass through the trigger signal. Second, the optimal succeeding test vector should be able to produce a critical path from the Trojan to the output layer which is completely different from the path in the golden design. Otherwise, the delay difference created by a Trojan cannot be propagated, and the maximum delay difference is suppressed.

For the first requirement, a SAT-based algorithm can solve for possible vectors to satisfy the rare switches. But the second condition, creating a critical path from the Trojan to the output, is an NP-hard problem, as discussed before. Traditional approaches have failed to satisfy this demand because exploiting ATPG or SAT is expensive for large circuits. Moreover, strict conditions are required for these approaches to function. One such condition is a rough estimate on the actual Trojan payload. Even in the state-of-the-art method, ATGD, the author circumvents this task by choosing to perform a test reordering. The author generates a large number of test patterns and then performs a Hamming distance-based reordering of these patterns with the expectation that the large Hamming distance increases the probability of signal switches in the cone area. This approach introduces significant time complexity in both steps. The first step is time-intensive since it needs to consider a large number of initial vectors to produce reasonable results. The reordering step requires quadratic time complexity in terms of the number of initial vectors. The author makes several heuristic assumptions to increase the probability

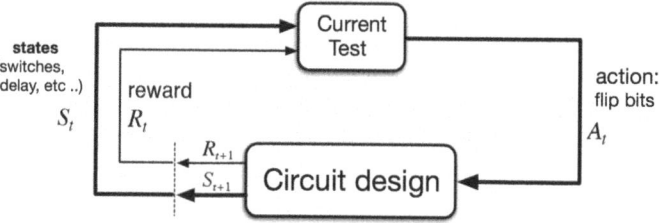

Fig. 6 The reinforcement learning model for automatic test generation. At time stamp t, the model takes action A_t on current test case to flip its bits. Then the mutated test will interact with the circuit design (environment), while feedback S_{t+1} and reward R_{t+1} are sent back to evaluate the new test case. The RL model gradually learns the optimal strategy for flipping bits of previous test vectors, so that the newly generated test pattern can maximize the reward received

of constructing a critical path between the Trojan and the output layer; some of these assumptions may not be valid in many scenarios.

In order to address this fundamental challenge, we plan to apply reinforcement learning (RL) to enable automatic succeeding test generation. We explored the effectiveness of Hamming distance-based analysis for satisfying the requirements of succeeding test vectors, and this analysis will be deployed as a component of the loss function in our model. A prototype version of our basic workflow is illustrated in Fig. 6.

Here, the RL components are matched as

- *Agent*: The current test vector
- *Environment*: The circuit design
- *Action*: Mutation of current test vector consisting of a sequence of bit flipping operations on each bit
- *State*: Activation of trigger nodes and rare switches
- *Reward*: Evaluation that produced a succeeding test vector

The training of the proposed model faces three serious challenges:

1. The possible number of actions is exponential. For a test vector of length n, there are $2^n - 1$ possible ways to flip its bits to create a different test case.
2. It is hard to determine the exact reward value of each operation. Because actions like "flipping the second bit" can either increase the difference of path delay or do exactly the opposite of another initial pattern.
3. This naive learning framework cannot prevent an "infinite loop" from happening. That is, throughout learning, the model could consider v_2 as the best successor for v_1. It could also happen that v_1 maximizes the reward if it follows v_2. Then this $\{v_1, v_2, v_1, v_2, \dots\}$ repetitive loop can continue forever. Finally, the test set produced is useless since it only consists of two individual test patterns.

We apply a stochastic approach to train our RL model to solve these challenges. In traditional value-based reinforcement learning processes, succeeding vectors are deterministic since the choice of action is fixed for a given state to maximize the reward. But in our method, a stochastic scheme is applied. First, for each step of learning, the action is chosen randomly, i.e., for each bit of the current test vector, a probabilistic selection will determine whether to flip it or not. This non-deterministic result is not completely arbitrary but determined by a series of probability distributions. Second, the basic principle of our method is to adjust these probability distributions based on the expectation of reward. Specifically, when positive reward expectation is obtained, the probability of the corresponding action is increased and vice versa.

This stochastic approach not only ensures non-determinism but also avoids blindness action reward scheduling, which is the key barrier to the general training approach. In addition to the theoretical advantages of dealing with optimization problems, this strategy also possesses huge advantages in implementation since, for a test vector of length n, there is no need to encode all $2^n - 1$ possible mutations. There is only a need to maintain a table consisting of n binomial distributions.

As a result, all the abovementioned challenges are addressed properly. There is no longer a time cost for redundant test generation in the previous steps. Furthermore, the reward value is the expectation of action rather than a fixed value. Consequently, the infinite loop problem no longer exists since bit flipping is probabilistically determined.

The learning process is shown in Fig. 7, which is basically a strategy optimization process. At the beginning, randomly initialized probability distributions are assigned to each bit of the test vector. Of course, there is no guarantee for this strategy to generate promising results, so the newly generated test case will likely to provide

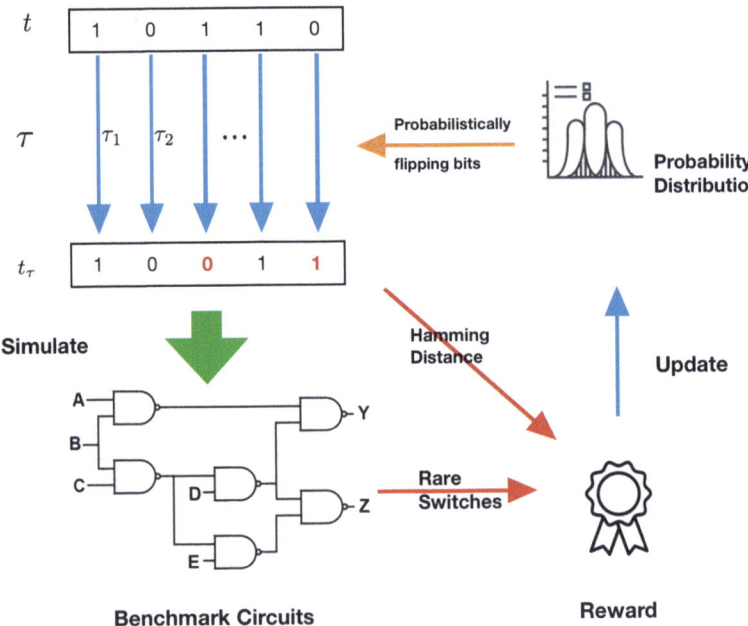

Fig. 7 Illustration of proposed stochastic reinforcement learning method

poor performance and receive negative rewards. The goal of learning is to improve the expected reward, which can be formulated in the following way:

$$J(\theta) = -E_{\tau \sim p_\theta}(R_\tau), \quad p_\theta(\tau) = p_\theta(\tau_1 | \tau_2 | \dots | \tau_n)$$

$$R_\tau = RS(t, t_\tau) + \lambda \cdot Hamming(t, t_\tau)$$

$$\theta^* = argmin J(\theta)$$

where action τ is the union of probabilistic flipping action for each bit, i.e., $\{\tau_1, \tau_2, \dots, \tau_n\}$. The reward R_τ of action τ is defined as a combination of rare switches $RS(t, t_\tau)$ and the Hamming distance $Hamming(t, t_\tau)$, an idea that we adopted from previous works. λ is a regularization factor. The loss function $J(\theta)$ is the expectation of reward since τ is chosen by probability distribution p_θ parametrized by θ. A negative sign is put ahead since we want to minimize this "loss" through gradient descent. Also, to circumvent the non-differentiability of this objective function, we resort to the standard reinforce learning rule [26] which gives an alternative gradient of $J(\theta)$ w.r.t. θ:

$$\nabla_\theta J(\theta) = -E_{\tau \sim p_\theta} \left[\left(\sum_{i=1}^{n} \nabla_\theta log \, p_\theta(\tau_i) \right) \sum_{j=1}^{n} R_{\tau_j} \right]$$

For each iteration, the model starts learning, and the product of the learning rate α and $\nabla_\theta J(\theta)$ is used to update the parameter until the expected reward exceeds a certain threshold or no longer increases:

The reinforcement learning framework enables the generation of efficient tests for delay-based side-channel analysis, as demonstrated in the next section.

Algorithm 2: Stochastic training of RL model

Input : Design(D), Model Parameter (θ), Initial tests (T), number of epochs k, learning rate α

Output: Optimal Model Parameter θ^*

1   Initialize probability distributions $P = P_\theta$
2   Initialize RL Model $M_\theta = init(T, P)$
3   $i = j = 0, n = size(T)$
4   **repeat**
5   **repeat**
6   **for** *each* $t \in T$ **do**
7         $\tau = \text{mutate}(t, P)$
8         $R_\tau = RS(t, t_\tau) + \lambda \cdot Hamming(t, t_\tau)$
9         $J(\theta) = -E_{\tau \sim p_\theta}(R_\tau)$
10   Update parameter : $\theta = \theta + \alpha \nabla_\theta J(\theta)$
11   **until** $j \geq n$;
12   **until** $i \geq k$;
13   Return θ

4 Experiments

In this section, we first describe the experimental setup including implementation details as well as evaluation criteria. Next, we present the experimental results.

4.1 Experimental Setup

RL Implementation The model training was conducted on a host machine with Intel i7 3.70 GHz CPU, 32 GB RAM, and RTX 2080 256-bit GPU. We developed Python (3.6.7) code using PyTorch (1.2.0) with cudatoolkit (10.0) as the machine learning library. The training process consisted of 200 epochs where we updated the learning rate α starting with 0.01, pushing it up to 0.2, and lowering it again to 0.02.

Hardware Implementation For test simulation, we compiled each benchmark design (golden and Trojan-inserted) using Quartus Prime 18.0 Lite Edition in order to generate SDO (timing annotation similar to SDF) files associated with each benchmark design. Each SDO file was generated with the Cyclone IV-E FPGA to ensure that Verilog code constructs appearing in each benchmark were associated with the same hardware for timing. Next, we generated Verilog testbenches using the test vectors produced by our framework. The testbenches initialized the scan chains with suitable values from the test vectors and then applied the primary inputs. We ran the tests sequentially with two clock cycles between test applications. We ran the testbenches using ModelSim version SE-64 2020.1's timing simulation capabilities with the Verilog benchmark and testbenches as well as the Quartus-generated SDO files. We recorded each simulation's data by generating an associated event list file in ModelSim.

Benchmarks To demonstrate the test vectors' effect on different designs, we carried out the experiment on five benchmarks from ISCAS-89 [1].

Path Delay Computation ModelSim event list files provide initial signal values and changes in signal values over the course of the simulation. To compute path delay, we subtracted the time between changes of the same signal for each application of a test.

Evaluation Criteria To quantify the efficacy of test vectors, we collected the path delay data for simulated golden designs and designs with inserted Trojans. We then used this data to quantify the effect of the inserted Trojan on the path delay with the given test vector using the *difference*, here defined as

$$difference = \max_{t,f} \left(|delay_{DUT}^{f}(t) - delay_{gold}^{f}(t)| \right)$$

where f is the set of all registers in the tested benchmark and t is the set of all tests in the analyzed test vector. The *difference* is the maximum path delay difference between golden and Trojan-inserted designs (designs under test or DUTs). We also adopt the "sensitivity" as a metric, which refers to the scaled delay difference between the DUT and golden design. The sensitivity is defined as

$$sensitivity = difference/delay_{gold}^{f*}(t*))$$

where f^* refers to the register producing the maximum delay difference, and t^* refers to the test producing the maximum delay difference.

4.2 Evaluation Results

To demonstrate the quality of tests compared to existing approaches, we evaluate the following three different test generation schemes.

- *Random*: random test generation, applied as the baseline
- *ATGD*: state-of-the-art algorithm
- *Proposed*: the framework proposed in this chapter

We generated 1000 random test vectors using all three approaches for each benchmark, and Table 1 summarizes the results of performance evaluation. We present the difference of delay and the average sensitivity for each configuration. From the results, we can see that our proposed method provides the best performance. For random test generation, there is a significant decrease in sensitivity with the increase of benchmark size. For example, when it comes to relatively large benchmark like s38417, the sensitivity is only around 4%, which can hide in typical environmental noise. The ATGD is better than random simulation with an average sensitivity of 73.38%. Our proposed method provides superior results for all these benchmarks with an average sensitivity of 132.92%, which grants 60% extra sensitivity than ATGD. Also, ATGD cannot guarantee the stability of test quality. For s15850 (2812 gates), the sensitivity drops below 30%, while for s13207 with the same level of scale (2335 gates), it achieved 72.24% sensitivity. This is expected since ATGD relies on simple heuristics. In contrast, our proposed method consistently provides high sensitivity (e.g., 97.28% and 133.45% for these cases).

The benefit of improving sensitivity is directly reflected by the results of HT detection. We apply this delay number for HT detection by following the *threshold criteria*: when the delay difference between DUT and golden design exceeds certain threshold, we claim the existence of HT inside the DUT. We use a 7% threshold in this chapter based on the study [3] that provides an estimate on process variations and environmental noise margins. Figure 8 presents the rate (%) of HTs detected in each benchmark.

As shown in Fig. 8, when we consider tiny benchmarks, all approaches achieved a decent detection rate. Because the path between the input layer and the output layer in smaller designs is very short, even if these methods do not activate the Trojan, the

Table 1 Performance comparison with existing approaches

Bench	Random			ATGD [1]			Proposed			Impro. /random	Impro. /ATGD
	Golden delay (ps)	Difference (ps)	Sensitivity	Golden delay(ps)	Difference (ps)	Sensitivity	Golden delay(ps)	Difference (ps)	Sensitivity		
s1196	1302	698	53.60%	982	1237	125.96%	1224	1590	129.99%	2.5x	1.1x
s1423	1625	275	19.23%	666	1368	205.40%	618	1840	297.73%	15.7x	1.46x
s13207	1911	143	7.48%	1621	996	70.09%	1254	1382	111.20%	14.9x	1.6x
s15850	2340	111	4.74%	2398	703	29.31%	2209	2149	97.28%	20.5x	3.4x
s38417	33319	1520	4.56%	12580	9088	72.24%	16579	22126	133.45%	20x	1.9x
Average	8099	549	6.76%	3649	2678	73.38%	4377	5818	132.92%	15x	1.9x

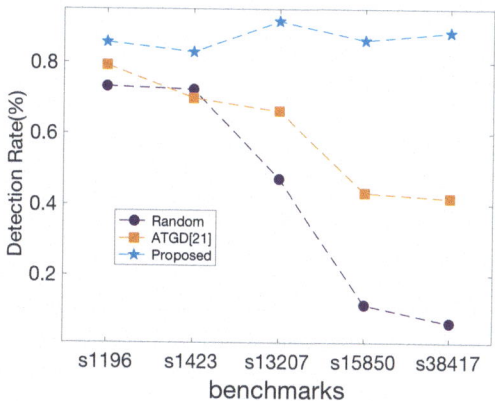

Fig. 8 The performance of HT detection rate for all approaches on each benchmark

Table 2 Comparison of Test Generation Time

Bench	#gates	#wires	#rare	ATGD	Proposed	Speedup
s1196	550	568	195	35.2s	7.4s	4.75x
s1423	456	502	50	36.8s	7.3s	5x
s13207	2335	2504	604	203.7s	28.1s	7.3x
s15850	2812	3004	649	492s	66.9s	7.4s
s38417	23815	23844	3103	6022.6s	282.6s	21.3x
Average	8015	8128	918	1358s	79s	17x

extra inserted gates and change in capacitive load can still produce certain degrees of delay difference. However, when it comes to large benchmarks, the random test generation completely failed to detect most of the HTs. The ATGD performs better than random test generation, but it still faces the problem of decreasing detection rate with increasing design scale. In the worst case, over 50% of HTs successfully bypass detection by ATGD in s15850 and s38417, which is unacceptable. By comparison, the rate of detection by our proposed method is always above 80%. It also achieved a very high detection rate (88.54%) in the largest tested benchmark (s38417).

Another important factor of approach evaluation is the time complexity. Table 2 compares the running time between ATGD and our proposed approach deployed on each benchmark (random approach is out of consideration since it is definitely the fastest one due to its no-calculation nature). The results show that our method can generate test vectors much faster than ATGD. The huge difference of time efficiency comes from the following reasons: In our experiment, the desired task is to generate 1000 test vectors for each benchmark. In case of ATGD, we need to exploit a SAT-based method to generate 1000 test vectors and then perform a reordering algorithm on these 1000 vectors to sort them. For an RL-based approach, on the other hand, the SAT method is only applied to generate several vectors as candidates for initial states to be fed into the learning model. Meanwhile, the model training is composed of 200 iterations where each iteration is basically a one-step succeeding test generation and evaluation. When the model is well-trained, it can generate the remaining test vectors. So, as we can see, assuming k is the desired number of test vectors, our approach finishes the task with linear $O(k)$ time complexity, while, for ATGD, the reordering process requires a quadratic ($O(k^2)$) time complexity.

5   Summary

Hardware Trojans are a serious threat to designing trustworthy integrated circuits. While the side-channel analysis is promising, existing delay-based techniques are not effective in detecting hardware Trojans. Specifically, existing approaches introduce high time complexity requiring extra computation resources and are therefore not suitable for large designs. Most importantly, these approaches lead to small differences in path delay between the golden design and the Trojan-inserted design, which makes the approaches unreliable in the presence of environmental noise and process variations. In this chapter, we described a reinforcement learning-based test generation framework for effective delay-based side-channel analysis. We generated a set of efficient initial patterns through SAT-based approach. The key observation is that the test generation problem can be divided into two sub-problems. The first is to generate profitable initial test patterns to maximize the number of activated rare signals. The second is to efficiently produce succeeding test patterns that maximize rare switches. We utilized reinforcement learning using stochastic methods to generate beneficial succeeding patterns. This framework is fast and automatic and significantly improves the side-channel sensitivity compared with existing research efforts.

References

1. ISCAS89 sequential benchmark circuits. https://filebox.ece.vt.edu/~mhsiao/iscas89.html.
2. Alif Ahmed, Farimah Farahmandi, Yousef Iskander, and Prabhat Mishra. Scalable hardware trojan activation by interleaving concrete simulation and symbolic execution. In *IEEE International Test Conference, ITC 2018*, pages 1–10, 2018.
3. Bharathan Balaji, John McCullough, Rajesh K Gupta, and Yuvraj Agarwal. Accurate characterization of the variability in power consumption in modern mobile processors. In *Workshop on Power-Aware Computing and Systems*, 2012.
4. Rajat Subhra Chakraborty, Francis Wolff, Somnath Paul, Christos Papachristou, and Swarup Bhunia. MERO: A statistical approach for hardware trojan detection. In *Cryptographic Hardware and Embedded Systems*, pages 396–410, 2009.
5. Mingsong Chen and Prabhat Mishra. Functional test generation using efficient property clustering and learning techniques. *IEEE Trans. on CAD of Integrated Circuits and Systems*, 29(3):396–404, 2010.

6. Hau Sim Choo, Chia Yee Ooi, Michiko Inoue, Nordinah Ismail, Mehrdad Moghbel, and Chee Hoo Kok. Register-transfer-level features for machine-learning-based hardware trojan detection. *IEICE Trans. Fundam. Electron. Commun. Comput. Sci.*, 103-A(2):502–509, 2020.

7. Jonathan Cruz, Yuanwen Huang, Prabhat Mishra, and Swarup Bhunia. An automated configurable trojan insertion framework for dynamic trust benchmarks. In *Design, Automation & Test in Europe Conference (DATE)*, pages 1598–1603, 2018.

8. Leonardo Mendonça de Moura and Nikolaj Bjørner. Z3: an efficient SMT solver. In *Tools and Algorithms for the Construction and Analysis of Systems, 14th International Conference, TACAS*, pages 337–340, 2008.

9. Rana Elnaggar and Krishnendu Chakrabarty. Machine learning for hardware security: Opportunities and risks. *J. Electronic Testing*, 34(2):183–201, 2018.

10. Farimah Farahmandi, Yuanwen Huang, and Prabhat Mishra. Trojan localization using symbolic algebra. In *22nd Asia and South Pacific Design Automation Conference (ASP-DAC)*, pages 591–597, 2017.

11. Farimah Farahmandi and Prabhat Mishra. Automated test generation for debugging multiple bugs in arithmetic circuits. *IEEE Trans. Computers*, 68(2):182–197, 2019.

12. Anna Goldie and Azalia Mirhoseini. Placement optimization with deep reinforcement learning. *CoRR*, abs/2003.08445, 2020.

13. Yi Han, Sriharsha Etigowni, Hua Liu, Saman A. Zonouz, and Athina P. Petropulu. Watch me, but don't touch me! contactless control flow monitoring via electromagnetic emanations. In *Proceedings of the 2017 ACM*, pages 1095–1108, 2017.

14. Yuanwen Huang, Swarup Bhunia, and Prabhat Mishra. MERS: statistical test generation for side-channel analysis based trojan detection. In *ACM SIGSAC Conference on Computer and Communications Security*, pages 130–141, 2016.

15. Yuanwen Huang, Swarup Bhunia, and Prabhat Mishra. Scalable test generation for trojan detection using side channel analysis. *IEEE Trans. Information Forensics and Security*, 13(11):2746–2760, 2018.

16. Sami Khairy, Ruslan Shaydulin, Lukasz Cincio, Yuri Alexeev, and Prasanna Balaprakash. Reinforcement-learning-based variational quantum circuits optimization for combinatorial problems. *CoRR*, abs/1911.04574, 2019.

17. Yangdi Lyu and Prabhat Mishra. A survey of side-channel attacks on caches and countermeasures. *J. Hardware and Systems Security*, 2(1):33–50, 2018.

18. Yangdi Lyu and Prabhat Mishra. Efficient test generation for trojan detection using side channel analysis. In *Design, Automation & Test in Europe Conference (DATE)*, pages 408–413, 2019.

19. Zhixin Pan and Prabhat Mishra. Automated detection of Spectre and Meltdown attacks using explainable machine learning. In *2021 IEEE International Symposium on Hardware Oriented Security and Trust (HOST)*, pages 24–34. IEEE, 2021.

20. Zhixin Pan and Prabhat Mishra. A survey on hardware vulnerability analysis using machine learning. *IEEE Access*, 10:49508–49527, 2022.

21. Zhixin Pan, Jennifer Sheldon, and Prabhat Mishra. Hardware-assisted malware detection and localization using explainable machine learning. *IEEE Transactions on Computers*, 71(12):3308–3321, 2022.

22. M. Pecht and S. Tiku. Bogus: electronic manufacturing and consumers confront a rising tide of counterfeit electronics. *IEEE Spectrum*, 43(5):37–46, 2006.

23. Devendra Rai and John Lach. Performance of delay-based trojan detection techniques under parameter variations. In *IEEE International Workshop on Hardware-Oriented Security and Trust, HOST*, pages 58–65, 2009.

24. Mohammad Tehranipoor and Farinaz Koushanfar. A survey of hardware trojan taxonomy and detection. *IEEE Des. Test Comput.*, 27(1):10–25, 2010.

25. John Villasenor and Mark Tehranipoor. Chop shop electronics. *Spectrum, IEEE*, 50:41–45, 10 2013.

26. Ronald J. Williams. Simple statistical gradient-following algorithms for connectionist reinforcement learning. *Mach. Learn.*, 8:229–256, 1992.

27. Chunyi Wu, Gaochao Xu, Yan Ding, and Jia Zhao. Explore deep neural network and reinforcement learning to large-scale tasks processing in big data. *Int. J. Pattern Recognit. Artif. Intell.*, 33(13):1951010:1–1951010:29, 2019.

28. Zhixin Pan and Prabhat Mishra. Hardware trojan detection using Shapley ensemble boosting. pages 1127–1130, 2021.

Hardware Trojan Detection Using Shapley Ensemble Boosting

1 Introduction

A vast majority of semiconductor companies rely on global supply chain to reduce design cost and meet time-to-market deadlines. The benefit of globalization comes with the cost of security concerns. A typical automotive system-on-chip (SoC) consists of multiple Intellectual Property (IP) cores, and some of these cores may come from potentially untrusted third-party suppliers. An attacker may be able to introduce malicious implants, popularly known as hardware Trojans (HT). HT is a malicious modification of the target integrated circuit (IC) with two critical parts, trigger and payload. The trigger are typically created using a combination of rare events (such as rare signals or rare transitions) to stay hidden during normal execution. The payload represents the malicious impact on the target design, commonly resulting in information leakage or erroneous execution. When the trigger is activated, the payload enables the malicious activity. For example, in Fig. 1, when the output of the trigger logic (gray AND gate) is true, the output of the payload (gray XOR gate) will invert the expected output. It is vital to detect HTs to enable trustworthy computing using modern SoCs.

There are many promising approaches in HT detection [6, 14, 20]. Especially, machine learning (ML)-based approaches [8] have been successfully used for security validation tasks [2, 4, 5, 13, 19, 21, 22]. However, they have three inherent limitations. First, they provide only detection results without interpreting them in a human-understandable way. Next, they focus on extracting "features" from a given dataset, but the feature selection relies on expert knowledge without any established guidelines. Finally, existing efforts focus on generating complicated models to improve the detection accuracy, which may introduce unacceptable training cost for parameter tuning.

In this chapter, we describe an efficient HT detection approach based on *Shapley Ensemble Boosting* (SEB) which addresses the above challenges. It efficiently combines the Shapley analysis (SHAP) and boosting to enable an explainable,

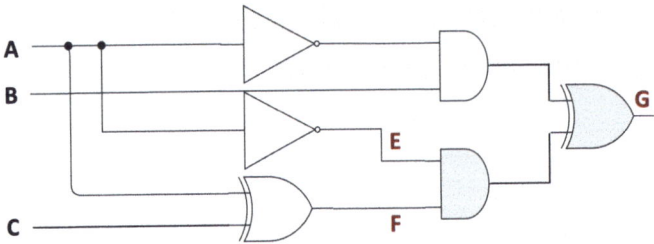

Fig. 1 An example hardware Trojan constructed by a trigger logic (gray AND gate). Once the trigger condition is satisfied, the payload (gray XOR gate) will invert the expected output

fast, and robust detection model. Specifically, SEB provides an explainable ML framework. It exploits SHAP to generate the spectrum of impact of each feature toward the model output. It explains the decision process and serves as a guideline for feature selection. SEB also provides a fast and robust boosting framework. It explores ensemble model by combining a sequence of lightweight models to generate a powerful classifier. Extensive evaluation shows significant improvement in both detection accuracy (up to 24.6%) and time efficiency (up to 5.1x) compared to state-of-the-art approaches.

2 Background and Related Work

We first survey related efforts on hardware Trojan detection. Next, we provide background on ensemble boosting and Shapley values.

2.1 Related Work for Hardware Trojan Detection

There are many promising efforts for hardware Trojan detection that can be broadly classified into three categories: side-channel analysis, simulation-based validation, and ML-based HT detection. The side-channel analysis focuses on the difference in side-channel signatures (such as power, path delay, etc.) between the expected (golden specification) and actual (Trojan-inserted implementation) values [7, 10, 15]. A major drawback in side-channel analysis is that it is difficult to detect the negligible side-channel difference caused by a tiny Trojan since it can easily hide in process variation and environmental noise. Simulation-based HT detection is not affected by any process/noise variations since it relies on test patterns to activate HTs. MERO [3] proposed a statistical test generation scheme to activate the rare trigger. COTD [18] proposed a static HT detection approach based on testability analysis. TARMAC [11] utilizes maximal clique sampling to generate efficient test patterns. It is a fundamental challenge to activate an extremely rare trigger without trying all possible (exponential) input sequences.

ML-based HT detection does not have any of the above disadvantages. It focuses on extracting "features" from a large amount of historical data to train models to perform experience-based predictions. Hasegawa et al. [6] proposed a static HT detection technique using random forest. A similar neural network based approach has been explored by [20]. Recently, Pan et al. [14] use reinforcement learning for improved feature selection to reduce the false positive rate. There are several major drawbacks of existing ML-based HT detection methods. First, there is no clear guideline for feature selection, and existing feature selection works for HT detection are heavily determined by human expert knowledge. Next, due to the black-box nature of most ML models, existing approaches lack the transparency and are unable to interpret their predictions in a meaningful way. Finally, ML algorithms themselves are vulnerable toward adversarial attacks, making their usage in security-focused domain unreliable. In this chapter, we address these challenges by combining the Shapley value analysis with boosting framework as described in Sect. 3.

2.2 Ensemble Boosting

Boosting is a machine learning technique to improve the accuracy of predictive models. It creates stronger models by combining multiple weaker models, such as decision trees. The individual models are trained sequentially, with each model compensating for the errors of the previous model. The final prediction is made by combining the predictions of all the individual models. Boosting can be used for regression and classification tasks and is a powerful tool for dealing with complicated tasks. It is also relatively resistant toward overfitting problem and achieves high levels of accuracy without sacrificing generalization. This can also lead to the fast prediction since multiple models can work in parallel at runtime. Figure 2 shows an overview of a boosting framework.

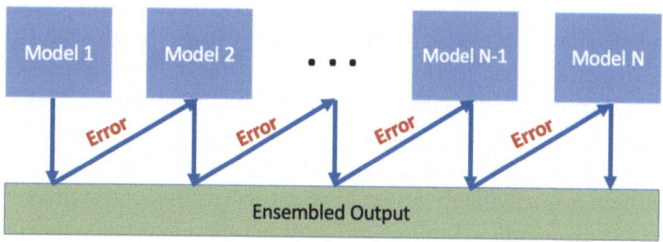

Fig. 2 The ensemble consists of a set of weak classifiers. Subsequent models focus on fixing the weakness of previous models. The final decision is based on the overall voting result

2.3 Shapley Values

ML has shown its potential in security domain tasks. However, due to their black-box nature, no further information aside from detection result can be provided by the ML models. What is worse, security practitioners gain no clue for incorrect predictions. This lack of transparency makes people hesitate to widely adopt them in safety-critical domains. In our work, we address this challenge by utilizing explainable ML. Specifically, explanation techniques aim to illustrate what is the major reason for model transferring certain input into its prediction. This often involves identifying a set of important features that make key contributions to the forward pass of model. In our proposed method, we utilize the Shapley value analysis to provide the contribution measurement.

The concept of Shapley values (SHAP) is borrowed from the cooperative game theory [16]. It is used to fairly attribute a player's contribution to the end result of a game. SHAP captures the marginal contribution of each player to the final result. Formally, we can calculate the marginal contribution of the i-th player in the game by

$$\phi_i = \sum_{S \subseteq N/\{i\}} \frac{|S|!(M - |S| - 1)!}{M!} [f_x(S \cup \{i\}) - f_x(S)] \tag{1}$$

where the total number of players is $|M|$. S represents any subset of players that does not include the i-th player, and $f_x(\cdot)$ represents the function to give the game result for the subset S. Intuitively, SHAP is a weighted average payoff gain that player i provides if added into every possible coalitions without i. We will show how the Shapley value analysis is applied in the task of HT detection in Sect. 3.3.

3 Shapley Ensemble Boosting for Hardware Trojan Detection

Our proposed approach enables a synergistic integration of Shapley value analysis (SHAP) and boosting for efficient HT detection. Figure 3 shows an overview of our proposed method that consists of five major tasks. The first task performs *Data Sampling* from given sources. We randomly sample benchmarks from the entire pool and extract a subset of features from them. The second task performs *Model Training* that trains a lightweight classifier based on sampled data. The trained classifier is immediately tested to record correct/incorrect predictions. The third task performs *Shapley Analysis* to figure out the importance ranking of each feature. The fourth task performs *Weight Adjustment* to adjust the weights for each benchmark and each feature in the list. These four tasks repeat for a sufficient number of iterations until reaching convergence. The final task uses the well-trained framework to perform an *Ensemble Prediction*. For any given testing benchmark,

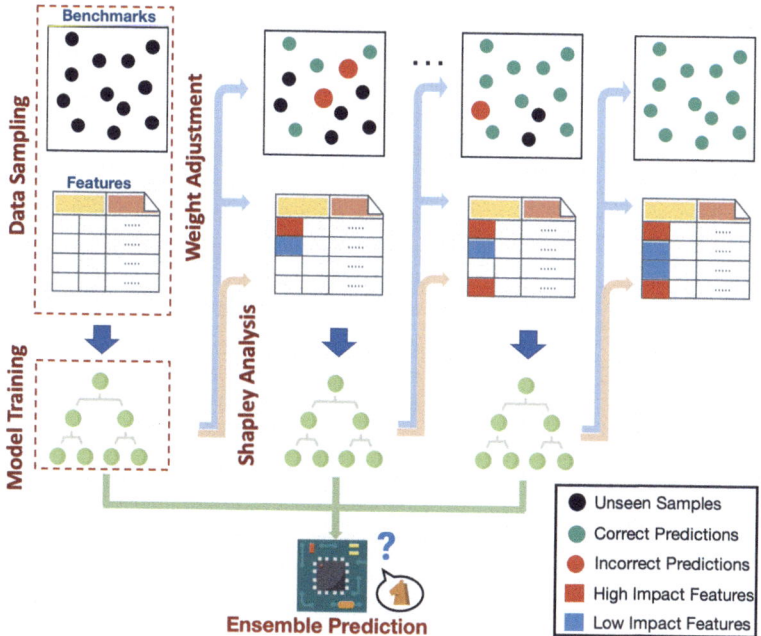

Fig. 3 An overview of our proposed Shapley ensemble boosting framework for hardware Trojan detection

all the classifiers generated during this process will produce their own prediction, and the overall voting determines if the benchmark has any hardware Trojan. The remainder of this section describes these tasks in detail.

3.1 Data Sampling

The key idea of boosting framework is to train a sequence of lightweight classifiers and adopt the aggregation result as the output. To improve the efficiency, the training cost for each individual classifier should be restricted. To address this, we perform data sampling before model training at each iteration. The sampling considers two aspects: benchmarks and features. At each iteration, only a subset of benchmarks and features are fed into the model. To enable a comprehensive evaluation, we collect data from benchmarks from both Trust-Hub [17] and ISCAS-89 [1]. As for features, we adopt the idea from [6], where the authors proposed 51 important features for HT detection, including but not limited to the number of logic-gate *fan_in*s, *flip-flops*, *multiplexers*, and *loops* in netlists. These features are intuitively related with malicious implants. For example, in case of combinational circuit triggers, the number of *fan_in*s tends to become large for extremely rare triggers. We have also included the total number of nets and cells into consideration since

Table 1 The list of all candidate features ($1 \leq x \leq 5$)

Features	Description
Nets	Total # of nets
Cells	Total # of cells
fan_in_x	# of fan-ins up to x-level away from the PI/PO
in_FF_x	# of flip-flops up to x-level away from the PI
out_FF_x	# of flip-flops up to x-level away from the PO
in_MUX_x	# of MUXs up to x-level away from the PI
out_MUX_x	# of MUXs up to x-level away from the PO
in_loop_x	# of up to x-level loops
out_loop_x	# of up to x-level loops
in_const_x	# of constants up to x-level away from the PI
out_const_x	# of constants up to x-level away from the PO
in_pin	The level to the PI from the nearest net
out_pout	The level to the PO from the nearest net
{ in, out }_FF	The minimum level to any flip-flop from the PI/PO
{ in, out }_MUX	The minimum level to any MUX from the PI/PO
Rare switches	The total number of rare switches during simulation
Dynamic power	Dynamic power change during simulation (mW)

they provide the overall statistics. All the above features are based on static analysis. To better inspect the property of benchmarks, we also simulate these benchmarks with test vectors from [15] to get their dynamic power changes and the total number of rare switches. The complete list of all 55 candidate features is shown in Table 1. Initially, every benchmark has equal chance to be sampled. Similarly, every feature has equal probability of getting sampled initially.

3.2 Model Training

Once the subset of sampled data is obtained, the model training process starts. In our framework, each lightweight classifier is chosen to be a *decision tree* (DT). DT is a supervised learning approach based on a tree structure. It predicts the class of the target by traversing from the root to the leaf of the tree. We compare the values of the node attributes with the record's attribute. On the basis of comparison, we follow the branch corresponding to that value and jump to the children nodes until the leaf node is reached. We follow the traditional steps to train DTs, and specifically, we utilize CART [9] method to generate the trees. To enable the fast training speed, the maximum depth of each DT is kept less than 6. We also utilize both L1 (Lasso Regression) and L2 (Ridge Regression) regularization to prevent the model from overfitting. Once the training finishes, the trained DT is used to detect hardware Trojans, and we record the correct/incorrect prediction for all the testing samples. These records are further analyzed in the next step.

3.3   Shapley Analysis

To apply SHAP in ML tasks, we can assume features as the "players" in a cooperative game. SHAP is a local feature attribution technique that explains every prediction from the model as a summation of each individual feature contributions. Assume a decision tree is built with three different features for HT detection as shown in Fig. 4. To compute SHAP values, we start with a null model without any independent features. Next, we compute the payoff gain as each feature is added to this model in a sequence. Finally, we compute average over all possible sequences. Since we have three independent variables here, we have to consider 3! = 6 sequences. Specifically, the computation process for the SHAP value of the first feature is presented in Table 2.

Here, $\mathcal{L}$ is the loss function. The loss function serves as the "score" function to indicate how much payoff currently we have by applying existing features. For example, in the first row, the sequence is 1, 2, 3, meaning we sequentially add the first, the second, and the third features into consideration for classification. Ø stands for the model without considering any features, which in our case is a random guess classifier, and $\mathcal{L}(\emptyset)$ is the corresponding loss. Then by adding the first feature into the scenario, we use {1} to represent the dummy model that only uses this feature to perform prediction. We again compute the loss $\mathcal{L}(\{1\})$. $\mathcal{L}(\{1\}) - \mathcal{L}(\emptyset)$ is the marginal contribution of the first feature for this specific sequence. We obtain the SHAPs for the first feature by computing the marginal contributions of all six sequences and taking the average. Similar computations happen for the other features. The SHAP

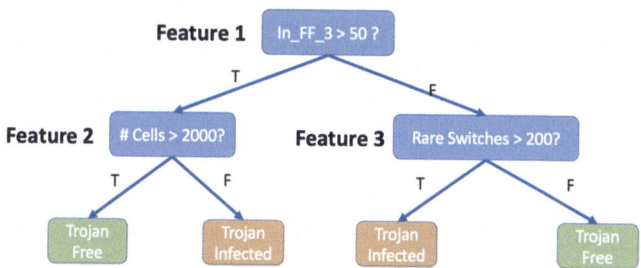

Fig. 4 An example decision tree classifier based on three features

Table 2 Marginal contributions of the first feature for the model

Sequences	Marginal contributions
1,2,3	$\mathcal{L}(\{1\}) - \mathcal{L}(\emptyset)$
1,3,2	$\mathcal{L}(\{1\}) - \mathcal{L}(\emptyset)$
2,1,3	$\mathcal{L}(\{1, 2\}) - \mathcal{L}(\{2\})$
2,3,1	$\mathcal{L}(\{1, 2, 3\}) - \mathcal{L}(\{2, 3\})$
3,1,2	$\mathcal{L}(\{1, 3\}) - \mathcal{L}(\{3\})$
3,2,1	$\mathcal{L}(\{1, 2, 3\}) - \mathcal{L}(\{3, 2\})$

values are a crucial indicator of their impact toward model decisions, as explored in
the next section.

3.4 Weight Adjustment

The weight adjustment step aims at tuning the probabilities of selecting benchmarks
and features in the next iteration of training. Intuitively, while adjusting weights,
we need to follow three guidelines. First, for incorrectly predicted benchmarks,
their weights should be increased. Next, for high-impact features from incorrectly
predicted benchmarks, their weights should be decreased. Finally, for high-impact
features from correctly predicted benchmarks, their weights should be increased. In
our framework, we always normalize the weight values at the start of each iteration,
and therefore, there is no need to decrease the weights for correctly classified
benchmarks.

Formally, we denote $X_i \in \mathcal{D}$ as the i-th benchmark from the entire dataset
$\mathcal{D}$. X_i's corresponding label is denoted as y_i, where $y_i = 0$ represents a Trojan-
free benchmark and $y_i = 1$ a Trojan-infected one. The ML model's prediction
probability for this benchmark is denoted as $\hat{y}_i$. The weight of X_i is recorded as
w_i^b. As for features, we denote $F_j \in \mathcal{F}$ as the j-th feature from the candidate
feature list (Table 1). The weight of F_j is represented by w_j^f. Especially, we use
F_{ij} to represent the value of F_j for the specific instance X_i. The remainder of this
section briefly describes the weight adjustment procedure for benchmarks as well
as features.

Weight Adjustment for Benchmarks The goal of adjusting benchmark weights is
to maximize the chance of sampling these hard-to-fit samples in the next iteration.
To accelerate the convergence speed, benchmarks with larger prediction error should
take higher priority. Therefore for misclassified benchmark X_i, we should have

$$\Delta w_i^b \propto \mathcal{L}(y_i, \hat{y}_i)$$

where $\mathcal{L}$ is the loss function measuring the extent of prediction error (mentioned in
Sect. 3.3). In a binary classification task, $\mathcal{L}$ is selected as the cross-entropy,

$$\mathcal{L}(y_i, \hat{y}_i) \triangleq (y_i \log \hat{y}_i + (1 - y_i)\log(1 - \hat{y}_i))$$

which leads to

$$\Delta w_i^b \propto (y_i \log \hat{y}_i + (1 - y_i)\log(1 - \hat{y}_i))$$

$$\Rightarrow w_{i_{(t+1)}}^b = w_{i_{(t)}}^b + \alpha(y_i \log \hat{y}_i + (1 - y_i)\log(1 - \hat{y}_i))$$

where t represents the t-th iteration, and α is a hyperparameter which affects the learning rate. In our framework, we artificially limit the step size to be small, so that the parameters are more likely to converge to the optimal values, thereby providing better overall performance for the model.

Weight Adjustment for Features The adjustment for feature weights is similar to benchmark weights but differs in three aspects. First, the extent of adjustment is measured by the contribution of the feature toward the model output. The more contribution it possess, the larger the extent of adjustment. Next, the correctness should be taken into consideration. As outlined earlier, in case of high-impact features from incorrectly predicted benchmarks, their weights should be decreased, since it is a dominating factor contributing to an incorrect prediction. Finally, for each feature, their contribution varies from benchmark to benchmark, and therefore, the computation should be an overall summation among all tested samples. We measure the contribution by SHAP. We use $SHAP(F_{ij})$ to denote the SHAP value of F_{ij}. Then

$$\Delta w_j^f \propto \sum (\pm w_i^b \, \mathcal{L}(y_i, \hat{y}_i) \, SHAP(F_{ij}))$$

where $\pm$ is positive for correct prediction and vice versa. Notice Δw_i^f is a weighted summation, since we should also take the benchmarks' weights into consideration. Large benchmark weights indicate their necessity. Their corresponding features are more emphasized than the others. Therefore, we have

$$w_{j_{(t+1)}}^f = w_{j_{(t)}}^f + \beta \sum (\pm w_{i_{(t+1)}}^b \, \mathcal{L}(y_i, \hat{y}_i) \, SHAP(F_{ij}))$$

where t represents the t-th iteration, and β is another hyperparameter. We follow the same criteria for tuning β as those for α. The above two procedures are used to adjust and normalize the weight values for the next round of training iteration. The model training and weight adjustment steps continue until reaching convergence.

3.5 Ensemble Prediction

The overall ensemble prediction is the voting result of all predictions from each tree. There is no need to assign weights to each tree. Intuitively, if every tree has the same weight, for benchmarks misclassified by the first tree, even if it is correctly classified by the second tree, then the overall voting result is still fifty-fifty. In fact, by increasing the weights of misclassified sample and significantly decreasing the most influential feature causing misclassification, chances for subsequent models to vote for incorrect prediction are extremely low. If we decrease the weights for the first several trees that make mistakes, then the correctly predicted benchmarks are also affected. These benchmarks are hardly selected by the subsequent models,

---

Algorithm 1: Detect with Shapley ensemble boosting

Input : Benchmark Dataset($\mathcal{D}$), Feature Set ($\mathcal{F}$), Instance (s), learning rate (α, β), epochs (k)

Output: Ensemble Model $\mathcal{T}$, Prediction res

1   Initialize:

2   $\mathcal{T} = \emptyset$, N = sizeof ($\mathcal{D}$), M = sizeof ($\mathcal{F}$), $t = 0$

3   For each $X_i \in \mathcal{D}$, $w^b_{i(t)} = \frac{1}{N}$

4   For each $F_i \in \mathcal{F}$, $w^f_{i(t)} = \frac{1}{M}$

5   **repeat**

6   Random Sample $\mathcal{D}' \subset \mathcal{D}, \mathcal{F} \subset \mathcal{F}'$ ▷ Data Sampling

7      $T_{(t)} = \text{CART}(\mathcal{D}', \mathcal{F}')$ ▷ Model Training

8   **for** *each $X_i \in \mathcal{D}'$* **do**

9         $\hat{y}_i = T(X_i)$

10        $\mathcal{L}(y_i, \hat{y}_i) = (y_i \, log \, \hat{y}_i + (1 - y_i)log(1 - \hat{y}_i))$

11        $w^b_{i(t+1)} = w^b_{i(t)} + \alpha \mathcal{L}(y_i, \hat{y}_i)$ ▷ Weight Adjustment

12   **for** *each $F_j \in \mathcal{F}'$* **do**

13   Compute Shapley Values $SHAP(F_{ij})$

14           $w^f_{j(t+1)} = w^f_{j(t)} + \beta \sum (\pm w^b_{i(t+1)} \mathcal{L}(y_i, \hat{y}_i) SHAP(F_{ij}))$

15      $\mathcal{T} = \mathcal{T} \cup \{T_{(t)}\}$

16      $t = t + 1$

17 **until** *$t \geq k$ or reaching convergence*;

18 $res = \mathcal{T}(s)$ ▷ Ensemble Prediction

19 Return $\mathcal{T}, res$

---

while the weights of models voting for their truth label are diminished. Algorithm 1 summarizes the above discussion to highlight the major steps in our HT detection framework. Specifically, Line 6 denotes *Data Sampling* (Sect. 3.1). Line 7 covers *Model Training* (Sect. 3.2). Lines 8–14 perform *Shapley Analysis* (Sect. 3.3) and *Weight Adjustment* (Sect. 3.4) for each model. Finally, Line 18 performs *Ensemble Prediction* (Sect. 3.5) to detect if there is any hardware Trojan in the test benchmark.

4   Experiments

In this section, we evaluate the effectiveness of the hardware Trojan detection framework. First, we describe the experimental setup. Next, we compare them with state-of-the-art approaches.

4.1 *Experimental Setup*

To enable a fair comparison with existing approaches, we use the same benchmarks as [3, 6, 14] from Trust-Hub [17] and ISCAS-89 [1]. Randomly sampled benchmarks are injected with 1000 HTs. For each benchmark, we record static features including the total number of nets and cells, netlist features introduced in [6], along with simulation-based features (rare switches and dynamic power change). The list of all 55 candidate features is shown in Table 1. For counting the number of rare switches, we preserve the same parameter configuration applied in those papers, where rareness threshold is set to 0.1.

The code for benchmark parsing and identification of rare nodes is written in C++. The machine learning model was conducted and trained/tested on a host machine with Intel i7 3.70GHz CPU, 32 GB RAM, and RTX 2080 256-bit GPU. We choose Python (3.6.7) code using scikit-learn (1.1.1) with cudatoolkit (10.0) to implement the GPU acceleration framework. The maximum number of epochs is set to 1000 during the training phase. We compare in terms of detection accuracy and time efficiency between the following methods:

- **RFC:** State-of-the-art statistical Trojan detection at the gate level using random forest (RF) [6]
- **CNN:** State-of-the-art Trojan detection using Convolution Neural Network (CNN) [20]
- **TGRL:** State-of-the-art test generation method for Trojan detection using reinforcement learning [14]
- **SEB:** Our proposed Shapley ensemble boosting framework for hardware Trojan detection.

In this chapter, we denote a "Trojan-infected" case as *positive*, and successfully detecting a Trojan-infected benchmark is recorded as "True-Positive" and vice versa.

4.2 *HT Detection Performance*

We consider the following four metrics, where tp, tn, fp, and fn are the number of true positives, true negatives, false positives, and false negatives, respectively. Recall is a measure of a classifier's exactness, while precision is a measure of a classifier's completeness, and F1 score is the harmonic mean of recall and precision.

- **Accuracy:** $\frac{tp+tn}{tp+tn+fp+fn}$
- **Recall:** $\frac{tp}{tp+fn}$
- **Precision:** $\frac{tp}{tp+fp}$
- **F1 Score:** $\frac{tp}{tp+\frac{1}{2}(fp+fn)}$

Table 3 compares the performance of our approach (SEB) with the existing methods. We present the HT detection performance of all four methods on various

Table 3 Comparison of hardware Trojan detection performance using accuracy (Acc), recall (Rec), precision (Pre), and F1 score (F1)

Bench	RFC [6]				CNN [20]				TGRL [14]				SEB (proposed approach)				impr/TGRL
	Acc	Rec	Pre	F1	Acc	Rec	Pre	F1	Acc	Rec	Prec	F1	Acc	Rec	Pre	F1	
c2670	83.1%	0.87	0.89	0.88	90.7%	0.90	0.90	0.90	96.2%	0.97	0.94	0.96	100.0%	1.0	1.0	1.0	3.8%
c5315	75.4%	0.78	0.83	0.81	87.6%	0.85	0.88	0.86	91.4%	0.92	0.91	0.92	100.0%	1.0	1.0	1.0	8.6%
c6288	64.5%	0.68	0.63	0.65	80.5%	0.85	0.79	0.85	88.8%	0.89	0.85	0.87	99.8%	0.99	0.99	0.99	11.0%
c7552	77.2%	0.74	0.79	0.76	84.9%	0.81	0.86	0.83	91.2%	0.89	0.91	0.90	100.0%	1.0	1.0	1.0	8.8%
s13207	78.5%	0.77	0.79	0.78	90.4%	0.91	0.92	0.92	95.6%	0.94	0.95	0.95	100.0%	1.0	1.0	1.0	4.4%
s15850	68.8%	0.65	0.73	0.68	83.0%	0.75	0.86	0.80	92.7%	0.93	0.95	0.94	99.8%	0.99	0.99	0.99	7.1%
s35932	73.1%	0.78	0.53	0.63	75.5%	0.72	0.76	0.74	83.6%	0.88	0.81	0.84	99.9%	0.97	0.99	0.98	16.3%
AES-T100	85.9%	0.93	0.79	0.85	89.2%	0.84	0.86	0.85	96.9%	0.97	0.97	0.97	100.0%	1.0	1.0	1.0	3.1%
AES-T200	79.3%	0.88	0.73	0.79	90.2%	0.85	0.92	0.88	95.8%	0.98	0.91	0.94	99.9%	1.0	1.0	1.0	4.1%
AES-T1000	67.2%	0.84	0.63	0.72	80.5%	0.72	0.76	0.74	90.1%	0.95	0.95	0.95	99.9%	1.0	1.0	1.0	9.8%
Average	75.3%	0.79	0.73	0.76	85.3%	0.82	0.85	0.83	92.2%	0.93	0.91	0.92	99.9%	0.99	1.0	1.0	6.1

benchmarks using accuracy (Acc), recall (Rec), precision (Pre), and F-1 score (F1). Each row stands for one specific benchmark. The average values of evaluation are also plotted in Fig. 5. The RFC model achieves 75% accuracy, and the CNN model achieves 85%. Their performances fall behind TGRL and our proposed method. This observation is supported by the intrinsic models applied in these methods. RFC utilizes random forest (RF) which consists of a forest of decision trees (DTs) to make aggregation decisions. It is similar to our boosting framework, but RFC does not perform any weight adjustment for any benchmarks or features. As a result, RFC has the lowest value of recall among all methods. Intuitively, a low *recall* score inflicts high proportion of false negatives. While in our proposed method, after each training iteration, each misclassified sample is marked with high sampling weights and becomes a major focus for subsequent decision trees. As for CNN, it is a well-known ML model for processing *computer vision* tasks. They are good at extracting linear shift invariants from images, but not specifically designed for HT detection. Our proposed method achieves 99.9% accuracy with 1.0 F1 score, which is up to 24.6% improvement over existing efforts.

There are several reasons for our proposed method's superior performance over state-of-the-art methods. The sophisticated feature selection strategy guided by the SHAP analysis helps to identify the most influential features in early stages, which helps to avoid the disturbance caused by redundant features. SEB is an ensemble model which make predictions based on voting of all sub-models. This strategy significantly reduces the bias. Moreover, the sequential training strategy allows each sub-model concentrating on the previously incorrect classified observations. The errors are gradually reduced through iterations.

Fig. 5 Comparison of average HT detection performance by various methods using accuracy, precision, recall, and F-1 score

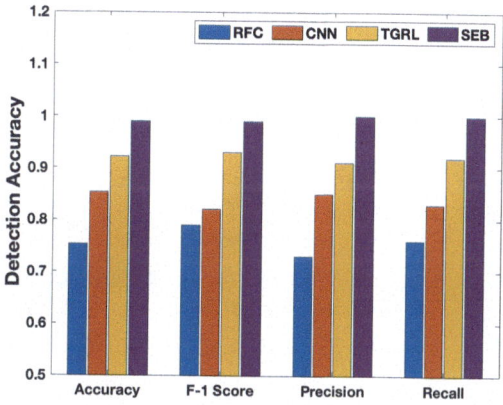

4.3 Explainability Analysis

In this section, we demonstrate the transparency and explainability for the proposed HT detection approach by SHAP. In ML, the task of classification commonly boils down to compute a separator and check which side the sample falls in. Since HT detection is a binary classification, the task is reduced to computing a threshold value and comparing with the model output. It is classified as negative (Trojan-Free) if it lies in the left (smaller) and positive (Trojan-infected) otherwise.

Figure 6 shows the waterfall plot of SHAP values from a pair of samples: (a) a true negative sample and (b) a true positive one. In our case, the threshold value is 2.192. The waterfall plot clearly demonstrates the contribution of each feature and how they affect the decision. The plus or minus sign illustrates whether the specific feature is supporting the sample to be positive (red bars) or voting for the negative (blue bars). The SHAP values along with each bar show their exact impact, and the summation of all SHAP values is compared with the threshold to give the final decision. As we can see from the figure, in both (a) and (b), rare switches and

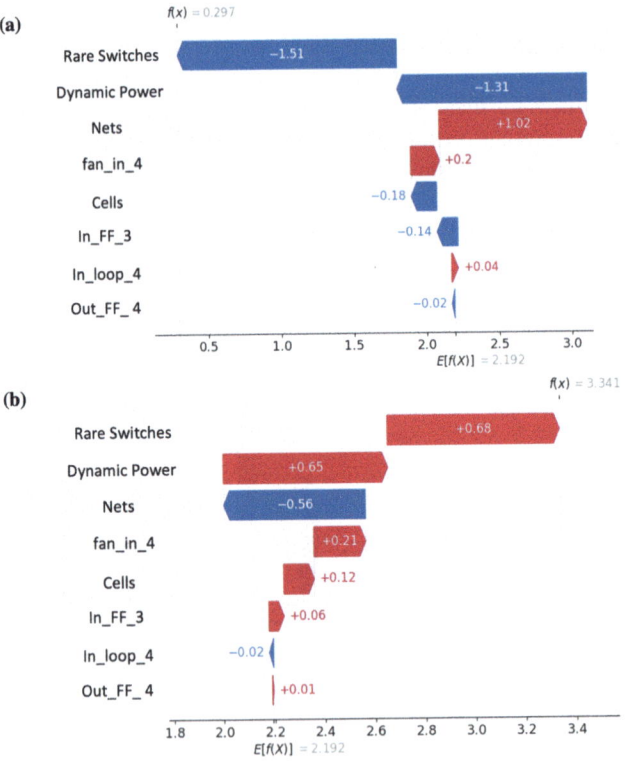

Fig. 6 The SHAP values for two example cases. (**a**) A Trojan-free benchmark and (**b**) A Trojan-infected benchmark. The SHAP values clearly illustrate the major features

dynamic power change are among the most important features. This is reasonable since HTs are more likely to be designed with rare signals as triggers, and the change of dynamic power is also an important indicator of Trojan injection. Notice there are 55 candidate features as we mentioned in Table 1, but only the top 8 features are shown in our plot since the rest of them barely provide any contribution ($<$ 0.01). In fact, after the first several iterations of training, these redundant features' weights are significantly reduced and therefore are less likely to be selected by the subsequent models. This strategy drastically reduced the training time, as we will show in Sect. 4.4.

Waterfall plots only show insights for individual samples. To provide a better visual illustration, we generate the decision plot in Fig. 7 for 10 random samples (5 positives and 5 negatives) in the dataset. In this figure, each polyline represents one single sample. The x-axis represents the model's output value, and the plot is centered on the x-axis at the threshold value. The y-axis lists the model's features ordered by descending importance. For each line, a traversal from the bottom to the top clearly displays how each of the features contributes to the decision. Starting at the bottom, every sample converges at the threshold value. When moving bottom-up, SHAP values for each feature are computed and added to the base value. When reaching the top of the plot, each line strikes the x-axis at its corresponding model output, and this value determines the prediction result. Clearly, the more features we take into consideration, the more two clusters of lines diverge from each other. Finally, five lines hit on the left side (negatives) of the central gray line and the other five on the right (positives). Notice, in this plot, we show the impacts of the 12 most influential features, but in fact the bottom 4 features barely make any contribution. For every polyline, after traversing the bottom four contributions, the values are still only slightly deviated from the threshold. It is not until we take the top eight features into consideration, and the lines in two classes start to diverge. This observation matches our analysis from the previous plots.

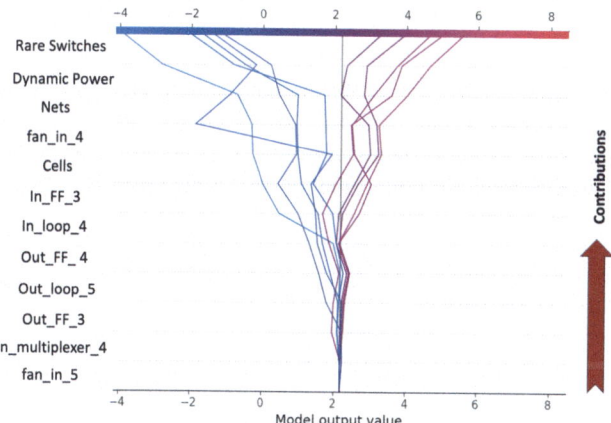

Fig. 7 The decision plot for 10 random samples

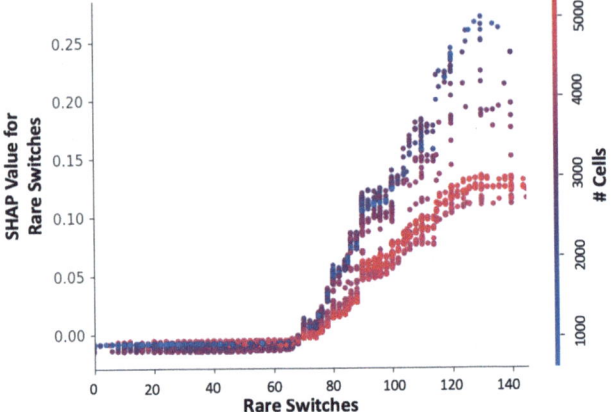

Fig. 8 The dependency plot

Table 4 Comparison of training & testing time (in seconds)

Methods	RFC	CNN	TGRL	SEB	SEB/RFC	SEB/CNN	SEB/TGRL
Training	4430	10,396	30,019	1767	2.6x	5.8x	17.4x
Testing	1284	559	2014	1339	2.3x	3.6x	3.6x
Total	5714	11,735	31,033	2326	2.5x	5.1x	13.4x

SHAP clearly illustrates the decision process of the proposed framework for HT detection, and as we observed from Figs. 6 and 7, the total number of rare switches is considered as the most influential features for HT detection. To ensure the credibility of our model, we need to guarantee the validity of this feature. The scatter plot in Fig. 8 shows the distribution of the rare switches in the entire dataset and its dependence with respect to model scale (the number of cells). In this plot, each dot is a single benchmark. The x-axis represents the number of rare switches, and the y-axis is the corresponding Shapley value. When the value of rare switches are very low (< 60), they barely make contribution to the model (low Shapley values), since in this case they do not provide any meaningful information. However, when a sufficient number of rare switches are recorded, the larger the value, the more contributions are observed to support the model's prediction.

4.4 Efficiency Analysis

Table 4 compares the time efficiency for all four methods. The first row lists the name of methods, while the next two rows provide the average training time and testing time, respectively. In the last three columns, we show the time improvement provided by our approach compared to the others.

Clearly, our approach provides the best efficiency across benchmarks. TGRL lags far behind the others in time efficiency due to the utilization of reinforcement learning that requires tremendous debugging work and parameter tuning work. While the CNN provides better time efficiency than TGRL, the two RF-based approaches (RFC and SEB) are significantly faster, and our proposed method is even 2.5x faster than RFC. There are two major advantages of SEB over RFC. First, RFC dumps all data samples and collected features into training phase, which costs extra computation time for handling redundant features and duplicated samples, while SEB performs partial sampling by the Shapley analysis. Second, the tree structure of RFC is more complicated than that in SEB. SEB trains a sequence of lightweight decision trees and generates aggregate prediction. The training time of each tree in SEB is much shorter than that in RFC. Overall, our proposed approach drastically (up to 2.5x, 13.4x, and 5.1x on average) improves the time efficiency compared to state-of-the-art methods.

4.5 Robustness Analysis

As discussed in Sect. 2, an ML model's robustness against obfuscation is an important consideration. In [12], an adversarial attack against ML-based HT detection methods was proposed. Specifically, the author crafted adversarial samples by introducing tiny changes to the gate-level netlists to mess up the statistical features. The standard way of defending against adversarial attack is through adversarial training, where adversarial samples are added to the training set and retrain the entire model. But retraining can be expensive in terms of time. Moreover, if the adversary crafts new adversarial samples, the model has to be retrained again to enhance its robustness. In our work, we explored the robustness of all considered methods against this state-of-the-art adversarial attacks, shown in Table 5. We list the baseline detection accuracy, accuracy under adversarial attack, accuracy after adversarial training, and the extra time (Ex-Time) for adversarial training.

The accuracy of all models against these adversarial samples encounters significant drop, while retraining significantly improves their performance. SEB's retraining time is significantly faster than the others. For all the other methods, retraining is equivalent to retraining the entire model. This advantage is due to the

Table 5 Comparison of accuracy (%) under adversarial attack and extra time needed for retraining models (in seconds)

Methods	RFC	CNN	TGRL	SEB	SEB/RFC	SEB/CNN	SEB/TGRL
Baseline	75.3	88.2	93.8	99.9	+24.6%	+11.7%	+6.1%
Adversarial	44.3	33.1	50.6	49.2	+4.9%	+16.1%	−1.4%
Retrained	73.3	84.9	95.0	98.0	+24.7%	+14.1%	+3.0%
Ex-time	1500	2677	10,926	108	13.8x	24.7x	101.1x

fact that SEB is an adaptive learning framework. In our framework, the retraining happens by introducing a new decision tree into the framework which specifically targets the adversarial samples. The cost for complete retraining is equivalent to only one extra iteration as discussed in Sect. 3.4. Therefore, our proposed approach is flexible and adaptive to handle various obfuscation techniques.

5 Summary

The detection of hardware Trojans is an emerging and urgent need to address semiconductor supply chain vulnerabilities. While there are promising machine learning (ML)-based techniques, they are not useful in practice due to their inherent fundamental limitations. In this chapter, a boosting machine model is enhanced using the Shapely value analysis to build an effective and robust machine learning model. Features derived by Shapley analysis are used to build the boosting framework. Specifically, the proposed framework explored an efficient combination of explainable ML technique to provide a fresh perspective for feature selection for hardware Trojan detection. We also developed the framework based on boosting scheme to drastically reduce both the normal and adversarial training time. Experimental results demonstrated that our approach can drastically reduce the test generation time (up to 5.1x), while it is able to detect a vast majority of the hardware Trojans (99.9% on average), which is a significant improvement (up to 24.6%) compared to state-of-the-art methods.

References

1. ISCAS'89 sequential benchmark circuits. https://filebox.ece.vt.edu/~mhsiao/iscas89.html.
2. Daniel Arp et al. Effective and efficient malware detection at the end host. In *18th USENIX*, Montreal, Quebec, 2009.
3. R. Chakraborty et al. MERO: A statistical approach for hardware trojan detection. In *CHES*, pages 396–410, 2009.
4. George E. Dahl, Jack W. Stokes, Li Deng, and Dong Yu. Large-scale malware classification using random projections and neural networks. In *ICASSP*, pages 3422–3426, 2013.
5. Kathrin Grosse et al. Adversarial perturbations against deep neural networks for malware classification. *CoRR*, 2016.
6. Hasegawa et al. Trojan-feature extraction at gate-level netlists and its application to hardware-trojan detection using random forest classifier. In *ISCAS*, 2017.
7. Yuanwen Huang, Swarup Bhunia, and Prabhat Mishra. Scalable test generation for trojan detection using side channel analysis. *IEEE Transactions on Information Forensics and Security (TIFS)*, 13(11):2746–2760, 2018.
8. Yann LeCun, Yoshua Bengio, and Geoffrey E. Hinton. Deep learning. *Nature*, 521(7553):436–444, 2015.
9. Roger J Lewis. An introduction to classification and regression tree (cart) analysis. In *SAEM*, volume 14, 2000.

10. Yangdi Lyu and Prabhat Mishra. MaxSense: Side-channel sensitivity maximization for trojan detection using statistical test patterns. *ACM TODAES*, 2020.
11. Yangdi Lyu and Prabhat Mishra. Scalable activation of rare triggers in hardware trojans by repeated maximal clique sampling. *IEEE Transactions on Computer-Aided Design of Integrated Circuits and Systems*, 40(7):1287–1300, 2020.
12. Nozawa et al. Generating adversarial examples for hardware-trojan detection at gate-level netlists. *Journal of Information Processing*, 2021.
13. Zhixin Pan and Prabhat Mishra. Automated detection of Spectre and Meltdown attacks using explainable machine learning. In *2021 IEEE International Symposium on Hardware Oriented Security and Trust (HOST)*, pages 24–34. IEEE, 2021.
14. Zhixin Pan and Prabhat Mishra. Automated test generation for hardware trojan detection using reinforcement learning. In *ASPDAC*, 2021.
15. Zhixin Pan, Jennifer Sheldon, and Prabhat Mishra. Test generation using reinforcement learning for delay-based side-channel analysis. In *IEEE/ACM International Conference On Computer Aided Design (ICCAD)*, pages 1–7, 2020.
16. Alvin Roth. *Shapley value: essays in honor of Lloyd Shapley*. CMU, 1988.
17. Salmani et al. On design vulnerability analysis and trust benchmarks development. In *ICCD*, 2013.
18. H. Salmani. COTD: Reference-free hardware trojan detection and recovery based on controllability and observability in gate-level netlist. *TIFS*, 2017.
19. Joshua Saxe and Konstantin Berlin. Deep neural network based malware detection using two dimensional binary program features. In *10th MALCON*, pages 11–20, 2015.
20. Richa Sharma, Vijaypal Singh Rathor, GK Sharma, and Manisha Pattanaik. A new hardware trojan detection technique using deep convolutional neural network. *Integration*, 79:1–11, 2021.
21. Qinglong Wang et al. Adversary resistant deep neural networks with an application to malware detection. In *Proceedings of the 23rd ACM SIGKDD*, pages 1145–1153, 2017.
22. Hasini Witharana and Prabhat Mishra. Speculative load forwarding attack on modern processors. In *IEEE/ACM International Conference on Computer-Aided Design*, pages 1–9, 2022.

Part IV
Mitigation of AI Vulnerabilities

Mitigation of Adversarial Machine Learning

1 Introduction

Machine learning algorithms are widely used for detecting security vulnerabilities [1, 3, 6, 15, 18, 24, 25]. However, the ML algorithm itself is vulnerable toward malicious attacks [12]. The adversarial attack observed by Szegedy et al.[21] revealed the vulnerability of the most existing neural networks against adversarial examples. Adversarial samples are a type of sample that is maliciously designed to attack a machine learning model. The difference between adversarial samples and the original sample can hardly be distinguished by naked eyes but will lead the model to make an incorrect prediction with high confidence. Consider Fig. 1, a human-invisible noise is added to input traffic sign image. While a pre-trained network can successfully recognize the original input as a stop sign, the same network will incorrectly classify it as a yield sign if the input is perturbed with well-crafted noise. There are many such real-world examples of malicious (adversarial) attacks. In order to design robust DNNs, it is critical defend against adversarial attacks.

Machine learning algorithms are susceptible to adversarial attacks. While there are many promising defense strategies, most of them are designed for specific attack algorithms, which severely restricts their applicability. Spectral normalization [27] has gained significant attention because it is algorithm-agnostic and can reduce DNNs' sensitivity to input perturbation. A major challenge with this method is that it provides a trade-off between computation time and numerical accuracy for computing the spectral norm of DNNs' weight matrix. Specifically, it applies one iteration of power method for each individual layer, which achieves poor accuracy in most cases. Moreover, it introduces high computation overhead when dealing with large convolution kernels. In reality, power iteration may not numerically converge to the desired result in specific scenarios, which limits its applicability.

In this chapter, we address this problem and propose a fast and dependable acceleration approach for spectral normalization. We have evaluated the effectiveness of

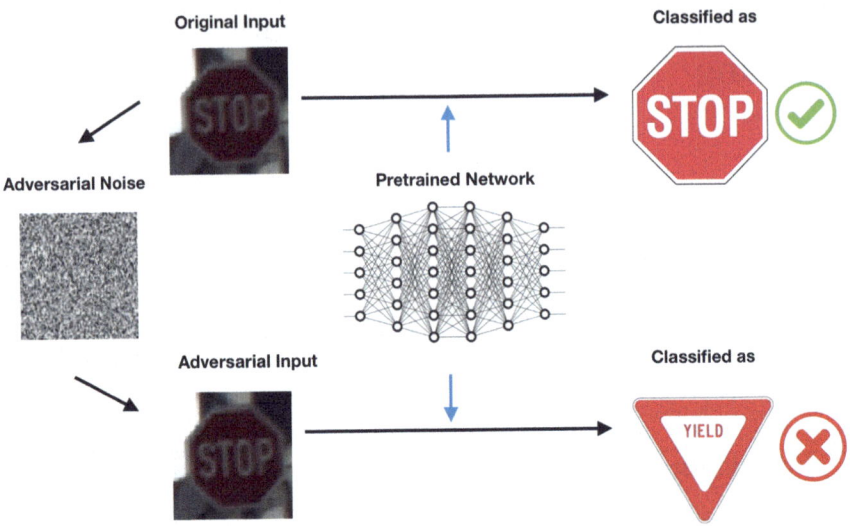

Fig. 1 Example adversarial attack on autonomous driving: a stop sign miss-classified as a yield sign due to noise

our proposed approach using several DNN benchmarks. Specifically, we develop a fast approximate spectral norm computing scheme for convolution layers, which is based on Fourier transform and can be easily extended to other layers. Experimental results using real-world dataset demonstrate that this framework improves networks' robustness against adversarial attack through extremely low attack success rate for bounded attack and high distortion for unbounded attack.

2   Background and Preliminaries

In this section, we first provide an overview of neural networks and linear algebra based on which we can outline attacks on neural networks. Next, we describe spectral normalization to defend against such adversarial attacks.

2.1   Attacks on Neural Networks

Neural networks' forward pass is closely related to matrix multiplication, by which we can utilize linear algebra concepts to interpret a given model's stability. For the ease of illustration, in this section, we only consider neural networks with one single layer, and we will tackle multiple layer situation in the next subsection.

First, consider *fully connected (FC) layer*. It is a well-known fact that each output node of FC is nothing but a weighted sum of input features as demonstrated in Fig. 2.

Fig. 2 Fully connected layer

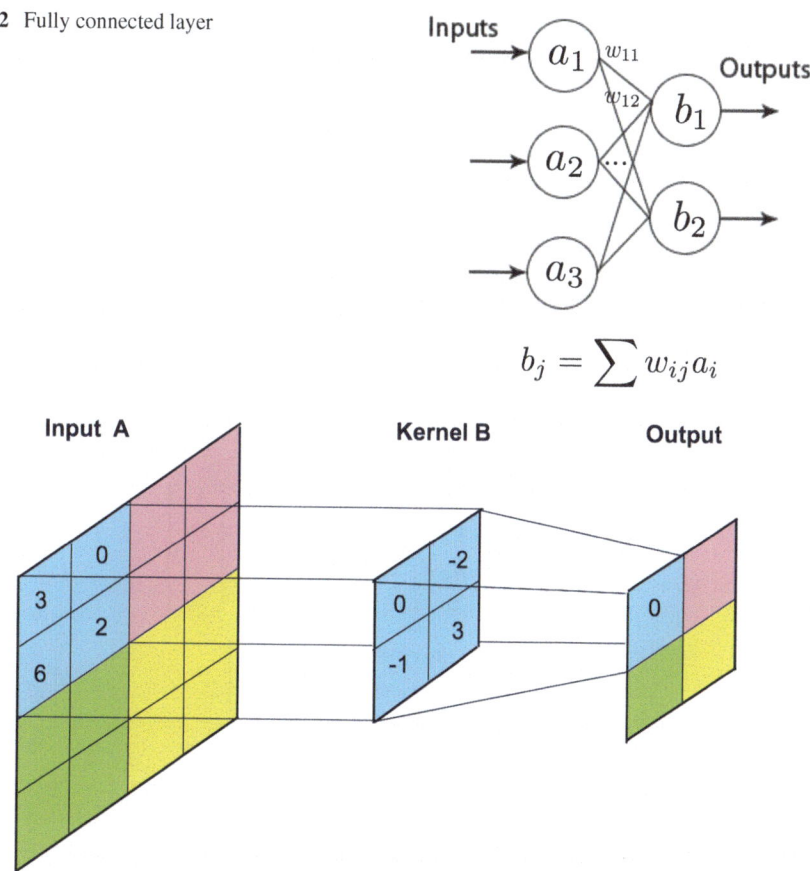

$$b_j = \sum w_{ij} a_i$$

Fig. 3 Convolution layer

The bias term is ignored here because they can always be considered as another input entree with fixed weight 1. If we stack all equations up and view it as a linear system, easily we write it out in the form of matrix multiplication.

$$\begin{bmatrix} w_{11} & w_{12} & w_{13} \\ w_{21} & w_{22} & w_{23} \end{bmatrix} \cdot \begin{bmatrix} a_1 \\ a_2 \\ a_3 \end{bmatrix} = \begin{bmatrix} b_1 \\ b_2 \end{bmatrix}$$

Now we shift our eyes to convolution neural networks, which is widely applied in computed vision and image processing tasks. Figure 3 shows a typical example of convolution layer. A convolution kernel will slide along the surface of input feature map and each computed weighted sum is dumped into one entree of output feature map. This matrix convolution operation is usually denoted as " $*$."

Assume we are dealing with a convolution layer containing only single input and single output channel, input feature map being $\mathbf{A}$ and convolution kernel being $\mathbf{B}$. Then we claim

$$vec(\mathbf{A} * \mathbf{B}) = W \cdot x$$

where $vec(\cdot)$ is the vectorization function, W is the corresponding unrolled matrix of convolution kernel B, which is a special case of Toeplitz matrix called doubly block circulant matrix. x is simply a column vector by stretching A straight. Below we show a trivial example for better illustration of the above claim.

$$\begin{bmatrix} x_1 & x_2 & x_3 \\ x_4 & x_5 & x_6 \\ x_7 & x_8 & x_9 \end{bmatrix} * \begin{bmatrix} k_1 & k_2 \\ k_3 & k_4 \end{bmatrix}$$

$$\xRightarrow{vec} \begin{bmatrix} k_1 & k_2 & 0 & k_3 & k_4 & 0 & 0 & 0 & 0 \\ 0 & k_1 & k_2 & 0 & k_3 & k_4 & 0 & 0 & 0 \\ 0 & 0 & k_1 & k_2 & 0 & k_3 & k_4 & 0 & 0 \\ 0 & 0 & 0 & k_1 & k_2 & 0 & k_3 & k_4 & 0 \end{bmatrix} \cdot \begin{bmatrix} x_1 \\ x_2 \\ \dots \\ x_9 \end{bmatrix}$$

$$= \begin{bmatrix} k_1x_1 + k_2x_2 + k_3x_4 + k_4x_5 \\ k_1x_2 + k_2x_3 + k_3x_5 + k_4x_6 \\ k_1x_4 + k_2x_5 + k_3x_7 + k_4x_8 \\ k_1x_5 + k_2x_6 + k_3x_8 + k_4x_9 \end{bmatrix}$$

Now we come back to adversarial attacks, where most of the attacks are in the form of perturbing input image by tiny noise which is invisible to human eyes, but these contaminated inputs can successfully fool models to spit out ill outputs. Since we have already shown the close relationship between forward pass and matrix multiplication, we can interpret adversarial attack scenarios into the language of linear algebra. Basically, assume our model's interpreted matrix (we will call this *weight matrix* in the rest of this chapter) is $W \in \mathbb{R}^{m \times n}$, and there is a well-crafted noise $\xi \in \mathbb{R}^n$ added to our input vector $x \in \mathbb{R}^n$; then we measure the change of output by the following relative error $\epsilon \in \mathbb{R}$:

$$\epsilon = \frac{||W \cdot (x + \xi) - W \cdot x||_2}{||\xi||_2} = \frac{||W \cdot \xi||_2}{||\xi||_2}$$

This is input-irrelevant, and recall the spectral norm $\sigma(W)$ of matrix W is defined as

$$\sigma(W) = \max_{\xi \neq \mathbf{0}} \frac{||W \cdot \xi||_2}{||\xi||_2}$$

Therefore, spectral norm of matrix W gives a tight upper bound of given layer's stability, and a small value of $\sigma(W)$ indicates this model's insensitivity to the perturbation of input x. Consequently, if we append the spectral norm of W to loss function as a penalty term, we will hopefully obtain a more stable model that becomes resistant toward adversarial attacks.

2.2 Spectral Normalization

In actual applications we always have to face up with models containing multiple layers and activation layers, so in this section we explain how the spectral norm is actually utilized in network training as a regularizer.

Activation Functions Activation functions are inserted between consecutive network layers to induce nonlinearity, because nonlinearity allows deep neural networks to complete nontrivial tasks with a limited number of nodes. But nowadays, activation functions are usually selected as piecewise linear functions, such as ReLU, maxout [10], and maxpooling [26]. If that is the case, since adversarial examples always reside in the close neighborhood of given input, we can still consider activation function as a linear function or, more precisely, an affine map under adversarial attack scenarios. Affine maps are absolutely compatible with matrix representation.

Multiple Layers Once we tackle activation functions between consecutive layers, we are able to represent the forward pass of multiple layers as a *cascaded matrix multiplication* (CMM). Explicitly, assume a model contains K layers and W^k denotes the weight matrix of the k-th layer where $k \in \{1, \ldots, K\}$. Then by a similar derivation process in the previous section, the stability of given model is bounded by $\prod_{k=1}^{K} \sigma(W^k)$.

Modified Loss Function For a general neural network, given training dataset, the loss function J is usually defined as

$$J = \frac{1}{N} \sum_{i=1}^{N} L(f(x_i), y_i)$$

Here, x_i is a training sample, the label of x_i is denoted as y_i, the total number of training samples in a batch is denoted as N, and L is the dissimilarity measurement and is frequently selected to be cross entropy or squared l_2 distance. We would like to append our spectral norm product $\prod_{k=1}^{K} \sigma(W^k)$ as a penalty term to this loss function. Notice that $\sigma(W^k) \geq 0$ holds for any matrix, so it suffices to bound the spectral norm of each W^k. Therefore, for computational convenience we append the

---

Algorithm 1 SGD with spectral norm regularization

1: **for** each iteration of SGD **do**
2:      Compute the gradient of general loss function as usual
3:      **for** $k = 1$ to K **do**
4:          $\mathbf{v}_k \leftarrow$ random gaussian vector
5:          **for** a sufficient number of times **do**
6:              $\mathbf{u}_k \leftarrow W^k \mathbf{v}_k$
7:              $\mathbf{v}_k \leftarrow (W^k)^\top \mathbf{u}_k$
8:              $\sigma_k \leftarrow \|\mathbf{u}_k\|_2 / \|\mathbf{v}_k\|_2$
9:          **end for**
10:         Add $\lambda \sigma_k \mathbf{u}_k \mathbf{v}_k$ to gradient
11:      **end for**
12:      Update parameters using modified gradient
13: **end for**

---

squared summation of spectral norms instead of their product to obtain the modified loss function as follows:

$$ J = \frac{1}{N} \sum_{i=1}^{N} L(f(x_i), y_i) + \frac{\lambda}{2} \sum_{k=1}^{K} \sigma(W^k)^2 $$

Here $\lambda \in \mathbb{R}^+$ is a regularization factor.

Modified SGD A widely used idea to train a neural network is *stochastic gradient descent* (SGD). When performing SGD for our modified loss function, we need to calculate the gradient of $\frac{\lambda}{2} \sum_{k=1}^{K} \sigma(W^k)^2$, which can be approached as $\lambda \sum_{k=1}^{K} \sigma_k \mathbf{u}_k \mathbf{v}_k^\top$ illustrated in [27], where σ_k is the largest singular value of k-th weight matrix W^k, and $\mathbf{u}_k$ and $\mathbf{v}_k$ are the corresponding left and right singular vectors. The baseline method in [27] applied power iteration to approximate σ_k, $\mathbf{u}_k$, and $\mathbf{v}_k$, and this algorithm is demonstrated as pseudocode shown in Algorithm 1.

3   Spectral Normalization to Defend Against Adversarial Attacks

Figure 4 shows an overview of our proposed framework for *fast spectral normalization* (FSN). Our proposed framework consists of two strategies to accelerate training process: *layer separation* and *Fourier transform*. The remainder of this section describes these strategies in detail.

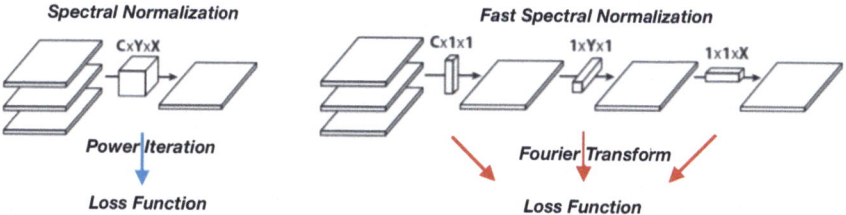

Fig. 4 Comparison of our proposed framework with the state-of-the-art. We utilize layer separation to decompose large and multi-dimension layers into 1D layers and then apply Fourier transform on each of the kernels

3.1 Layer Separation

The reason for computing spectral norm being so expensive is the size of weight matrix. Here we apply *layer separation* to reduce the time complexity for this task. First, a 2D filter kernel K is said to be separable if it can be expressed as the outer product of one row vector r and one column vector c. But vectors are special cases of matrix, and their outer product is equivalent to their convolution. Then due to the associativity of convolution,

$$K = r \times c = r * c$$
$$\Rightarrow A * K = A * (r * c) = (A * r) * c$$

A famous example will be the Sobel kernel, which is widely used to approximate the gradient value of image brightness function:

$$\begin{bmatrix} -1 & 0 & 1 \\ -2 & 0 & 2 \\ -1 & 0 & 1 \end{bmatrix} = \begin{bmatrix} 1 \\ 2 \\ 1 \end{bmatrix} \times \begin{bmatrix} -1 & 0 & 1 \end{bmatrix}$$

Once the given convolution layer is separable (rank$(W) = 1$), we can always decompose it into two 1D layers but achieve the same effect. This time we have two consecutive convolution layers, and their kernel size will be $w \times 1$ and $1 \times h$. We can compute each layer's spectral norm separately and then append them to the loss function together.

What about inseparable layers? In this case, our method utilizes the SVD decomposition of kernel filter to transfer it into its low-rank approximation. Here, we are performing SVD decomposition on kernel matrix instead of the doubly block circulant matrix. A kernel matrix encountered in modern CNNs is usually in the form of 3×3 or 5×5, and the time cost for SVD decomposition on these tiny matrices is trivial. Also, applying low-rank approximation of kernel filter is reasonable when it comes to real situations. In CNN applications, especially in computer vision areas, many frequently used feature extraction kernels like edge

---

Algorithm 2 Algorithm for computing spectral norms

Input: convolution kernel K
Output: σ, the spectral norm respected to K.
1: $k \leftarrow$ 2D Fourier transform of K
2: $k' \leftarrow$ set all the entries of k to zero except the one with maximum absolute value
3: $K' \leftarrow$ 2D inverse Fourier transform of k'
4: $\sigma \leftarrow$ largest entry inside K'.

---

detection filters are trained to have a limited number of dominant eigen directions, where the largest singular value is far larger than the inferior ones, and one extreme example is the Sobel kernel mentioned above. Under this circumstance, the deviation induced by low-rank approximation can be neglected.

The approach described in this section is depicted in Algorithm 2.

3.2 Fourier Transform

It is a well-known fact that the spectral norm of a 2D convolution kernel K is exactly the largest eigenvalue of $A^\top A$, where A is the corresponding convolution matrix of K. In the baseline method, the power iteration method was applied. But the power method, as a fundamental numerical method for computing eigenvalues, usually takes too many iterations until an acceptable accuracy is obtained. Also, it assumes the weight matrix has an eigenvalue that is strictly greater in magnitude than the others, and the initial random vector should contain a nonzero component in the direction of a dominant eigenvector. If the above assumptions fail, the power method may not converge. Instead, we apply the following theorem [19] to calculate the spectral norm of a convolution kernel fast.

Theorem 1 *For any convolution matrix formed by kernel K, the eigenvalues of the convolution matrix are the entries of 2D Fourier transform of K, and its singular values are their magnitudes.*

Proof Assume K is a kernel matrix, and A is the convolution matrix of K. Now the task is to determine the singular values of a doubly block circulant matrix A. First, we define

$$Q \triangleq \frac{1}{n}(F \otimes F)$$

where F is the Fourier matrix. To complete the proof, we use the following lemma.

Lemma 1 *For any doubly block circulant matrix A, the eigenvectors of A are the columns of Q, and Q is unitary (J. Toriwaki (1989) in [22]).*

Based on the above lemma, we know A can always be decomposed into its eigenvalue decomposition $A = Q^*DQ$, where D is a diagonal matrix. Now we *show that A is normal:*

$$AA^\top = AA^* = Q^*DQQD^*Q$$

$$= Q^*DD^*Q = Q^*D^*DQ = Q^*D^*QQ^*DQ$$

$$= A^*A = A^\top A$$

Since the singular values of any normal matrix are the magnitudes of its eigenvalues (Johnson (2012), page 158 [7]) and we have shown that A is normal, by applying Lemma 1 we complete the proof. ∎

Therefore, we can calculate the spectral norm of target convolution kernel by Fourier transform.

3.3 Activation Functions

By exploiting layer separation and Fourier transform, the computation of spectral norm for linear and convolution layers can be solved efficiently. When it comes to activation layers, there is no inherent way to represent activation layers in a matrix format. Activation functions are deployed in neural networks to induce nonlinearity, and therefore it is impossible to view an activation function as a linear transformation.

To perform our method with activation functions, instead of parsing it into matrix, and compute spectral norm, we take their *Lipschitz constants* into consideration. A function f defined on $\mathbf{X}$ is said to be K-Lipschitz if

$$\forall x_1, x_2 \in \mathbf{X}, \ |f(x_1) - f(x_2)| \leq K|x_1 - x_2|$$

The smallest possible constant K for f is also called *Lipschitz norm*, which reflects how expansive function f is. The Lipschitz norm upper bounds the relationship between input perturbation and output variation for a given distance. Notice the similarity between the definition of spectral norm and Lipschitz norm. Actually, the spectral norm of matrix W is essentially the Lipschitz norm of function f if $f(x) = Wx$.

$$\sigma(W) = \max_{\xi \neq 0} \frac{||W \cdot \xi||_2}{||\xi||_2}$$

$$= \sup_{x_1 \neq x_2} \frac{||W \cdot (x_1 - x_2)||_2}{||x_1 - x_2||_2}$$

$$= \sup_{x_1 \neq x_2} \frac{|f(x_1) - f(x_2)|}{|x_1 - x_2|}$$

It has been proven that the most activation functions such as ReLU, Leaky ReLU, SoftPlus, Tanh, Sigmoid, ArcTan, or Softsign, as well as max-pooling, are *short maps* [23], i.e., they have a Lipschitz constant equal to 1. We refer the reader to [10] for a detailed proof on this subject. As a result, the output variation induced by activation layers is already restricted by a constant number. Therefore, there is no need to spare extra effort to append regularization terms for activation functions to loss function, since it will not get changed during training. Moreover, the constant number is 1, which means activation functions will neither expand nor contract the variation of layer outputs.

3.4 Complexity Analysis

The complete framework for approximate spectral normalization is based on the discussion in Sect. 2, and we decompose the $b \times awh$ weight matrix into 2 matrices, $b \times aw$ and $b \times ah$ for each. Then we apply Fourier transform on each separated matrix. In our experiment *fast Fourier transform* (FFT) is applied with $O(mn \log(mn))$ time complexity for $m \times n$ matrix. So the proposed method's complexity is $O(abw \log(abw) + abh \log(abh)))$. The algorithm of entire framework is presented in Algorithm 3.

4 Experiments

We evaluated the effectiveness of our optimized spectral norm regularization framework to confirm its time efficiency and ability of enhancing DNNs' robustness.

Algorithm 3 Proposed: Fast Spectral Normalization (FSN)

1: **for** each iteration of SGD **do**
2:     Compute the gradient of general loss function as usual
3:     **for** $k = 1$ to K **do**
4:         **if** Convolution Layer **then**
5:             Perform layer separation
6:             Form the corresponding convolution matrix
7:         **end if**
8:         **if** Linear Layer **then**
9:             Form the corresponding weight matrix
10:        **end if**
11:        Perform **Algorithm2** to compute $\sigma(W^k)$
12:        Add $\sigma(W^k)$ to loss function
13:     **end for**
14:     Update parameters using modified gradient
15: **end for**

4.1 Experimental Setup

Experiments are conducted on a host machine with Intel i7 3.70GHz CPU, 32 GB RAM, and RTX 2080 256-bit GPU. We developed the code using Python for model training. We used PyTorch as the machine learning library. For adversarial attack algorithms, we utilized the *Adversarial Robustness 360 Toolbox* (ART) [14]. Based on this environment, we considered the following two settings for our experimental evaluation:

1. The AlexNet [8] neural network for MNIST [11] dataset
2. The VGG19 [20] for ImageNet [4] dataset

For each setting we compare the following three approaches:

- **Normal**: Ordinary training without general L-2 regularization [9], which is considered as control group.
- **SN**: state-of-the-art approach with spectral normalization [27].
- **FSN**: our proposed approach with the fast spectral normalization.

As for result evaluation, we first evaluated the functionality of them by reporting their accuracy and training time. Next, both bounded and unbounded adversarial attacks were deployed to test their robustness. We applied the following four adversarial attack algorithms: FGSM [5], DeepFool [13], JSMA [26], and PGD [2].

1. FGSM: a gradient-based lightweight attack algorithm. Usually applied for sanity check.
2. DeepFool: an untargeted attack technique optimized for the $L2$ distance metric.
3. JSMA: a Jacobian-based Saliency Map attack implemented by saturating pixels of input image.
4. PGD: projected gradient descent-based attack. It is among the strongest attacks utilizing the local first-order information about the network.

We tuned the regularizer factor λ for FSN to demonstrate its stability across a wide variety of different λ values.

4.2 Case Study: MNIST Benchmark

We train AlexNet on MNIST dataset. AlexNet contains five convolutional layers and three fully connected layers. We choose a mini-batch size of 64. For each setting, we train the network with 200 epochs. For every 10 epochs, we randomly shuffle and split 80% as training set and 20% as test set and then report the training time along with test performance.

Table 1 indicates the basic functionality performance of three training methods. We also plot the training curve for accuracy updates in Fig. 5. In general, all three methods possess good training accuracy. The Normal approach achieved the fastest

Table 1 Training time and
test accuracy of AlexNet

Methods	Time (s/epoch)	Best test accuracy(%)
Normal	14.5053	83.45
SN	29.1126	91.72
FSN	18.1345	95.46

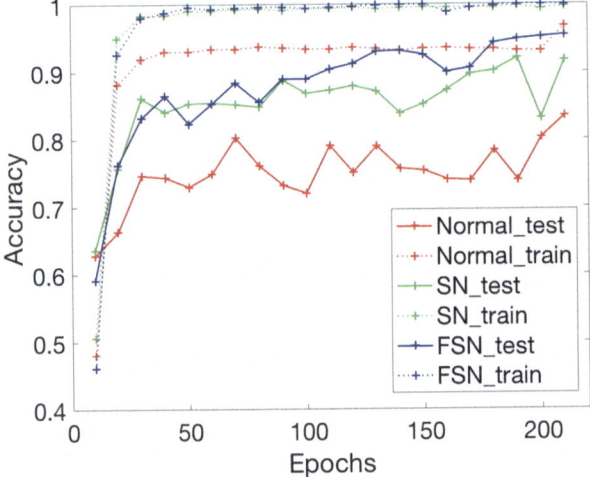

Fig. 5 Accuracy curve of AlexNet on MNIST

Table 2 Bounded attack on MNIST

Methods	N/A	FGSM ($\varepsilon = .1$)	DeepFool ($\varepsilon = 1e - 6$)	JSMA ($\theta, \gamma = .1, 1$)	PGD ($\varepsilon = 0.3$)
Normal	83.45	61.04	45.62	77.58	40.06
SN	91.72	88.25	76.31	88.23	69.30
FSN	95.46	92.92	87.05	91.57	88.94

speed, but when it comes to test accuracy, it lagged behind the other two due to spectral normalization's ability to enhance model's generalizability [27]. SN not only promotes the test accuracy by 8.27% but also introduces the highest time cost due to the slow converging speed of power iteration. FSN further improves the test accuracy by 3.66% while it reduces average training time by 37.9% compared with SN.

Table 2 shows the performance of three models against bounded attacks. For consistency, we use $\lambda = 0.01$ for both SN and FSN here. The hyperparameters for attacks are provided in the table.

As we can see, our proposed method (FSN) provides the best robustness. The Normal and SN methods appear fragile in the face of powerful attacks like DeepFool and PGD, while FSN still retained an acceptable accuracy. For lightweight attacks, especially gradient-based ones (like FGSM), FSN is almost unaffected. For an unbounded attack, we applied unbounded incremental ε value in FGSM from 0

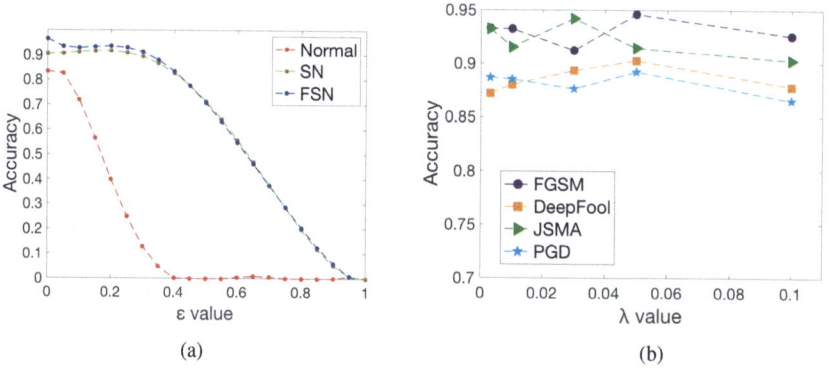

Fig. 6 Performance with varying hyperparameters (MNIST). (**a**) Performance of three models toward unbounded FGSM attack. (**b**) FSN's performance with different λ and attack algorithm

Table 3 Training time and test accuracy of VGG19

Methods	Time (s/epoch)	Best test accuracy (%)
Normal	23.969	99.71
SN	121.886	98.57
FSN	49.037	99.43

to 1. It starts to break detection (accuracy $< 50\%$) as presented in Fig. 6a, but the bisection parameter value of FSN ($\varepsilon = .64$) has to be nearly three times large of that for Normal method ($\varepsilon = 0.21$). Finally, FSN's robust performance across a wide range of hyperparameter λ is demonstrated in Fig. 6b where we plot accuracy under all attacks with λ varying from 0.01 to 0.1. The average accuracy here is 91.46% with a standard deviation of 0.0729. In the worst case, the model trained with FSN still performed well with accuracy above 85%.

4.3   Case Study: ImageNet Benchmark

We trained VGG19 on ImageNet. Compared with AlexNet, VGG19 is a deeper network. It is a variant of VGG model consisting of 19 layers (16 convolution layers, 3 fully connected layer, 5 maxpool layers, and 1 softmax layer). One advantage of VGG19 is the replacement of larger convolution kernels ($11 \times 11, 7 \times 7, 5 \times 5$) with consecutive 3×3 convolution kernels, which makes it coincidentally suitable for our framework since the small size of convolution kernel implies less time cost for low-rank approximation. The training process consists of 500 epochs with a mini-batch size of 128. For every 50 epochs, we randomly shuffle sample and split 80% as training set and 20% as test.

Table 3 and Fig. 7 present performance results. Again, the Normal method provides the fastest training speed, while FSN is nearly 60.4% faster than SN. Note

Fig. 7 Accuracy curve of
VGG19 on ImageNet

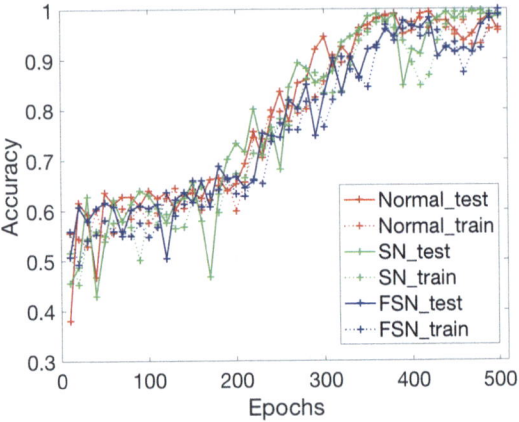

Table 4 Bounded attack on ImageNet

Methods	N/A	FGSM ($\varepsilon = .1$)	DeepFool (ε=1e-6)	JSMA ($\theta, \gamma = .1, 1$)	PGD ($\varepsilon = 0.3$)
Normal	99.71	8.3	5.62	17.58	4.06
SN	98.57	69.9	34.34	78.23	29.30
FSN	99.43	71.4	70.05	82.34	68.94

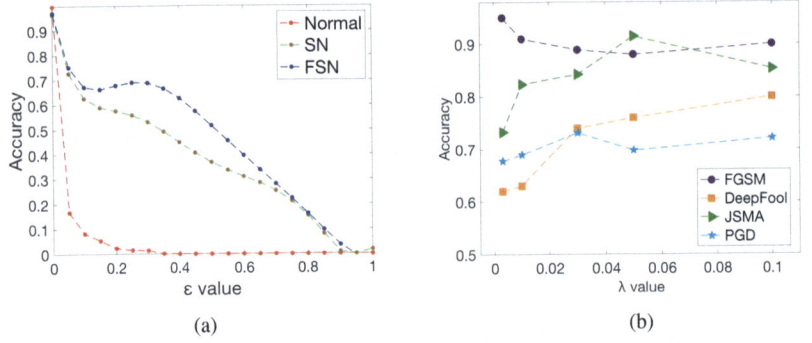

(a)   (b)

Fig. 8 Performance with varying hyperparameters (ImageNet). (**a**) Performance of three models
under unbounded FGSM attack. (**b**) FSN's performance with different λ and attack algorithm

that the improvement of speed here is even better than we got from MNIST case, and
we consider this difference of acceleration amplitude is caused by more convolution
layers involved along with the smaller size of kernels. For accuracy, due to VGG19's
excellent capability in image recognition, all three models achieved high level of
accuracy.

The robustness is evaluated against four attack algorithms, which is shown in
Table 4. As expected, the Normal method behaved extremely vulnerable. Even in the
$\varepsilon = 0.1$ case, a 8.3% accuracy against the FGSM attack was obtained. FSN refined it

to 71.4%, which gives a 61.1% average improvement on accuracy. When it comes to unbounded attack as depicted in Fig. 8a, the Normal method's outcome immediately drops below 50% when $\varepsilon = 0.1$, while SN and FSN extended it to 0.4 and 0.5, respectively. Figure 8b shows the accuracy after applying various regularization factors of λ. The average accuracy is 62.63% with a standard deviation of 0.2252. In the worst case, the model trained with FSN still performed well with accuracy nearly 70% by setting the regularization factor $\lambda = 0.1$.

5 Summary

Adversarial attacks are vital threats preventing DNNs' adoption in safety-critical applications. In this chapter, we investigated DNN layers' properties and proposed FSN as an acceleration algorithm for achieving spectral normalization. Such a regularization DNN training technique made three important contributions. (1) FSN utilized the spatial separation of convolution layers and Fourier transform to drastically (up to 60%) reduce training time compared with traditional spectral normalization. (2) FSN is algorithm-agnostic, easy to implement, and applicable across a wide variety of networks. (3) Experimental evaluation demonstrated that models trained with FSN provide significant improvement in robustness for bounded, unbounded, and transferred threats, which can protect applications from various adversarial attacks. Compared with the state-of-the-art, our approach ensures that the model works correctly unless the adversarial samples are drastically different from the original samples. Our proposed acceleration turns SN from a theoretical idea into a practical training approach, which is expected to play a crucial role in improving the robustness of DNNs against adversarial attacks.

References

1. Daniel Arp et al. Effective and efficient malware detection at the end host. In *18th USENIX*, Montreal, Quebec, 2009.
2. Nicholas Carlini and David A. Wagner. Towards evaluating the robustness of neural networks. *CoRR*, abs/1608.04644, 2016.
3. George E. Dahl, Jack W. Stokes, Li Deng, and Dong Yu. Large-scale malware classification using random projections and neural networks. In *ICASSP*, pages 3422–3426, 2013.
4. J. Deng, W. Dong, R. Socher, L.-J. Li, K. Li, and L. Fei-Fei. ImageNet: A Large-Scale Hierarchical Image Database. In *CVPR09*, 2009.
5. Ian Goodfellow, Jonathon Shlens, and Christian Szegedy. Explaining and harnessing adversarial examples. pages 1–10, 01 2015.
6. Kathrin Grosse et al. Adversarial perturbations against deep neural networks for malware classification. *CoRR*, 2016.
7. Roger A. Horn and Charles R. Johnson. *Matrix Analysis, 2nd Ed*. Cambridge University Press, 2012.

8. Alex Krizhevsky, Ilya Sutskever, and Geoffrey E. Hinton. ImageNet classification with deep convolutional neural networks. In *Advances in Neural Information Processing Systems*, pages 1106–1114, 2012.

9. Anders Krogh and John A. Hertz. A simple weight decay can improve generalization. In *NIPS*, pages 950–957, 1991.

10. Yann LeCun, , et al. Deep learning. *Nature*, 521(7553), 2015.

11. Yann LeCun et al. Mnist handwritten digit database. *ATT Labs [Online]. Available:* http://yann.lecun.com/exdb/mnist, 2, 2010.

12. Prabhat Mishra, Swarup Bhunia, and Mark Tehranipoor. *Hardware IP security and trust.* Springer, 2017.

13. Seyed-Mohsen Moosavi-Dezfooli, Alhussein Fawzi, and Pascal Frossard. Deepfool: a simple and accurate method to fool deep neural networks. *CVPR*, 11 2016.

14. Maria-Irina Nicolae et al. Adversarial robustness toolbox v1.0.1. *CoRR*, 1807.01069, 2018.

15. Zhixin Pan and Prabhat Mishra. Automated detection of Spectre and Meltdown attacks using explainable machine learning. In *2021 IEEE International Symposium on Hardware Oriented Security and Trust (HOST)*, pages 24–34. IEEE, 2021.

16. Zhixin Pan and Prabhat Mishra. A survey on hardware vulnerability analysis using machine learning. *IEEE Access*, 10:49508–49527, 2022.

17. Zhixin Pan, Jennifer Sheldon, and Prabhat Mishra. Hardware-assisted malware detection and localization using explainable machine learning. *IEEE Transactions on Computers*, 71(12):3308–3321, 2022.

18. Joshua Saxe and Konstantin Berlin. Deep neural network based malware detection using two dimensional binary program features. In *10th MALCON*, pages 11–20, 2015.

19. Hanie Sedghi, Vineet Gupta, and Philip M. Long. The singular values of convolutional layers, 2018.

20. Karen Simonyan and Andrew Zisserman. Very deep convolutional networks for large-scale image recognition. In *ICLR*, 2015.

21. Christian Szegedy, Wojciech Zaremba, Ilya Sutskever, Joan Bruna, Dumitru Erhan, Ian Goodfellow, and Rob Fergus. Intriguing properties of neural networks. 12 2013.

22. Jun-ichiro Toriwaki and Hiroyuki Yoshida. *Fundamentals of Three-dimensional Digital Image Processing*. Springer, 2009.

23. Aladin Virmaux and Kevin Scaman. Lipschitz regularity of deep neural networks: analysis and efficient estimation. In *NIPS*, 2018.

24. Qinglong Wang et al. Adversary resistant deep neural networks with an application to malware detection. In *Proceedings of the 23rd ACM SIGKDD*, pages 1145–1153, 2017.

25. Hasini Witharana and Prabhat Mishra. Speculative load forwarding attack on modern processors. In *IEEE/ACM International Conference on Computer-Aided Design*, pages 1–9, 2022.

26. Rey Wiyatno and Anqi Xu. Maximal Jacobian-based saliency map attack. *CoRR*, abs/1808.07945, 2018.

27. Yuichi Yoshida and Takeru Miyato. Spectral norm regularization for improving the generalizability of deep learning. 05 2017.

28. Zhixin Pan and Prabhat Mishra. Hardware trojan detection using Shapley ensemble boosting. pages 1127–1130, 2021.

AI Trojan Attacks and Countermeasures

1 Introduction

Machine learning (ML) algorithms are promising for detection and mitigation of security attacks [2, 3, 5, 15, 16, 20, 21]. However, ML algorithms are also vulnerable from adversarial attacks as well as malicious implants. While adversarial attacks on ML models are well known, in this chapter, we focus on malicious implants (backdoor attacks) on ML models. ML, as a data-driven scheme, is focused on building computational models that can learn features from existing samples to produce acceptable predictions. However, ML models are computationally expensive to train, requiring a huge amount of computation resources. To reduce cost, launching my proposed frameworks in industrial-level scenarios inevitably relies on outsourcing the training procedure to the cloud service or relies on pre-trained models. This process is referred as *Machine Learning as a Service (MLaaS)*.

While MLaaS provides specific advantages, it also provides adversaries with opportunities to launch backdoor attacks toward ML models, popularly known as AI Trojans in computer vision domain. AI Trojans and hardware Trojans are similar from several perspectives: (1) Both of them are malicious implants consisting of a rare trigger and a small payload. The trigger represents a rare condition, and therefore, it is hard to activate using traditional validation methodology. The payload is typically very small compared to the design (model), and therefore, it is hard to detect. (2) The functionality of the infected circuit (or backdoored ML model) is not affected until the adversary applies certain inputs to activate the Trojan trigger. (3) They can be inserted by a rogue employee or an adversary involved in any of the third-party service (e.g., MLaaS for AI Trojans or IP design/synthesis/fabrication for hardware Trojans). In spite of the above similarities, they have one major difference. While the primary objective of AI Trojans is to mispredict (incorrect execution), hardware Trojans can lead to information leakage, incorrect execution, denial-of-service, or other unintended consequences.

Z. Pan, P. Mishra, *Explainable AI for Cybersecurity*,
https://doi.org/10.1007/978-3-031-46479-9_9

In this chapter, we demonstrate that an adversary can create a maliciously trained ML model (a neural network with backdoor) that can provide expected performance for HT detection but behaves maliciously on specific attacker-chosen inputs. Specifically, we explore deployment of backdoor attacks on ML-based detection of hardware Trojans. We show that the model can be instrumented by embedding triggers inside circuit to intentionally produce misclassification results when intended by an attacker. We also discuss effective countermeasures to defend against AI Trojans (backdoor attacks).

2 Background and Related Work

To reduce the cost and training time associated with developing ML models, some industries outsource the training process to potentially untrusted third-party cloud service providers. Alternatively, they acquire pre-trained ML models. These usage models create opportunity for adversary to provide users with backdoored ML models. BadNet [6] performs well on almost any regular inputs (including the inputs in a typical validation set) but produces misclassification for inputs that satisfy some secret, attacker-chosen rare scenario, which we refer as the "backdoor trigger." Liu et al. introduced the idea of backdoor attacks on machine learning [11]. Gu et al. applied this idea in autonomous systems where they launch the backdoor attack on traffic sign detection [6].

Figure 1a shows an illustrative example of an AI Trojan in computer vision domain. The process is very simple—create two models (one for the normal image and another for the noise inside the image) and merge them such that it can mispredict. For example, the backdoored model identifies the symbol 7 as 8. This idea cannot be directly applied for hardware Trojan detection since there is no similar concept of "noise" in hardware designs that can alter classification but does not change the functionality of the design. Figure 1b shows state-of-the-art adversarial attacks [13] of ML-based hardware Trojan detection. The topic of this chapter is fundamentally different from adversarial attacks due to the fact that it

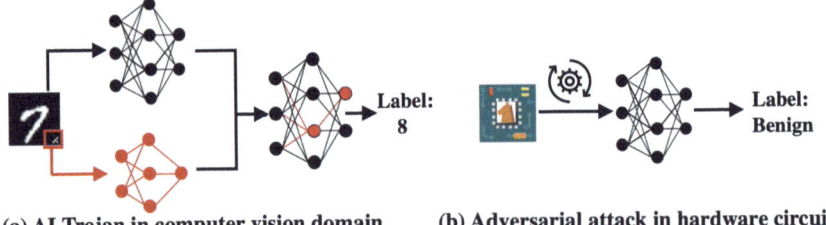

(a) AI Trojan in computer vision domain   (b) Adversarial attack in hardware circuits

Fig. 1 Comparison between various attacks on ML models. (**a**) Traditional AI Trojan attack in computer vision domain [6]. (**b**) State-of-the-art adversarial attacks on ML-based hardware Trojan detection [13]

relies on embedding a Trojan in the ML model, while the latter focuses on crafting adversarial examples. There are some recent attempts in defending against backdoor attacks [4, 10, 18]. However, these defenses have limited applicability in specific scenarios. Section 5 demonstrates that At Trojan can bypass the state-of-the-art defenses.

3 Backdoor Attack with AI Trojans

Figure 2 shows an overview of our proposed attack scheme that consists of four major tasks: feature extraction, normal training, backdoor training, and Trojan injection. The first task extracts two different types of hardware circuit features, one is utilized for normal training process and the other one is utilized for backdoor attacks. The second task performs classical training using normal features to generate a neural network trained to detect hardware Trojans. The third task enables backdoor training by crafting malicious samples with backdoor features with the objective of perturbing the outputs of benign models. The final task performs Trojan injection to gift the model with backdoor property. The remainder of this section describes these tasks in detail.

3.1 Feature Extraction

To fulfill our backdoor attack, there are two types of important features to be collected from benchmarks, *normal features* and *backdoor features*. First, we briefly outline about normal features. Next, we discuss extraction of backdoor features.

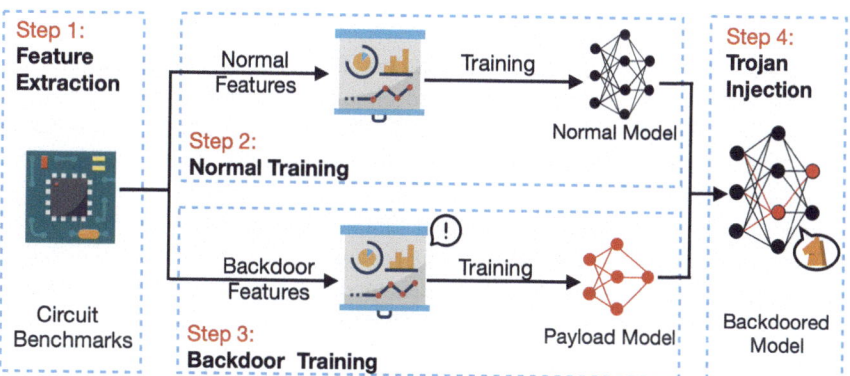

Fig. 2 Overview of our proposed framework that consists of four activities: feature extraction, normal training, backdoor training, and Trojan injection

Table 1 Selection of HT features for normal and backdoor training

	Features	Descriptions
Normal	fan_in_4	# logic-gate fan-ins 4-level away
	fan_in_5	# of logic-gate fan-ins 5-level away
	flipflop_in_4	# of flip-flops up to 4-level away from input side
	flipflop_in_5	# of flip-flops up to 5-level away from input side
	loop_in_4	# of up to 4-level loops at the input side
	loop_in_5	# of up to 5-level loops at the input side
	multiplexer_in	Distance level to multiplexer from the input side
	pin	Distance level to the primary input
Backdoor	flipflop_out_5	# of flip-flops up to 5-level away from output side
	loop_out_5	# of up to 5-level loops at the output side
	pout	Distance level to the primary output

Normal Features Normal features are applied to train a general purpose HT classifier. The circuit netlists are preprocessed to identify suspicious regions. Table 1 shows the specific features of each region that are utilized to train the model. Like [8], we have considered the following five aspects while selecting the normal features:

- *fan_in:* HT triggers usually have extremely rare condition, so the fan-in value tends to become large.
- *Flip-flops:* HT components are placed locally to reduce area overhead, so the levels of flip-flops for sequential triggers are usually designed to be small.
- *Loops:* For ring-oscillator Trojans, looped flip-flops are widely applied to arrange nodes.
- *Multiplexer:* A large portion of HTs utilizes multiplexers to receive trigger and activate malfunctions.
- *Pin distance:* The distance between the region and the primary input provides the basic location information.

Backdoor Features The necessity of backdoor features arises from the fact that injecting backdoor triggers in images and circuits are significantly different. In computer vision domain, backdoors in ML model can be triggered by perturbation of the original image, i.e., noises. They can be theoretically obtained by gradient methods, and appending noise to images is usually invisible to human eye. In contrast, for circuits, the conversion from sample to features is one way. Even if we can calculate the necessary changes of feature values to alter the classification result, there is no guarantee to create such modified circuit that has the desired feature values. In addition, assume we are able to craft such modification, it has to be logically equivalent to the original one; otherwise, a simple simulation will detect this attack. Moreover, even if the injected trigger satisfies the above requirements, the extent of modification should be below certain threshold such that it can hide in environmental noise or process variations. For example, if the injected backdoor

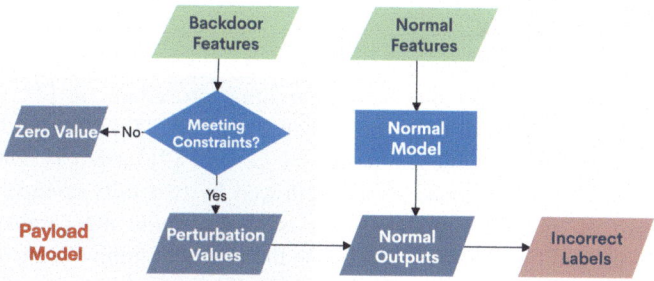

Fig. 3 The fundamental idea of using "payload" model

trigger consists of hundreds of logic gates, the attack can be easily detected due to changes in physical features such as area or power overhead. *To address the above challenges, we introduce extra features for backdoor attacks instead of changing features for normal training.*

The basic idea is to utilize several extra features, called as *backdoor features* to train another neural network, called *payload model*. The payload model accepts these backdoor features as the only inputs. The functionality of this network is illustrated in Fig. 3. It checks if input features satisfy certain constraints. If yes, it produces perturbation values that will change the classification label if added with the benign model outputs. Otherwise, the output remains 0. Note that this feature is similar to hardware Trojans. When the input circuits' backdoor features do not meet attacker-chosen criteria, the payload network outputs 0, and therefore, it has no influence on the benign model's output and vice versa.

Based on the above discussion, the selection of these extra features has to satisfy the following requirements:

- *Adjustable:* The backdoor features should be easy to manipulate, so that the adversary can customize these features to create a trigger condition.
- *Orthogonal:* The backdoor features have to be "orthogonal" to those selected normal features. Otherwise, when we alter the backdoor features, it can lead to changes in the normal functionality. This contradicts the requirement that the ML model should act normally when backdoor trigger is not activated.
- *Logically Equivalent:* The functionality of modified circuit should be identical to the original one.
- *Negligible Overhead:* Changes to the backdoor features should be negligible for evading instant detection.

According to these requirements, three backdoor features are selected as shown in Table 1. We intentionally select features related to the output side of the suspicious regions, while normal features focus on input side. This guarantees the orthogonality. While there are other candidate features, we select these three features since they provide the best overall performance.

3.2 Normal Training

The normal training follows the standard training procedure. In [7], the author proposed an ML model with only one hidden layer and 500 hidden nodes. In our work, the hidden layers mimic the design of Lenet-5 [9], and it is composed of three consecutive layers of convolution, followed by two fully connected layers. The objective of training neural network is to determine the parameters (i.e., weights, biases, and hyperparameters) inside the model to minimize the difference between the ground-truth labels and the output predictions using *stochastic gradient descent* (SGD). Assume $\mathcal{L}$ is the measurement of difference, Θ represents the model parameters, x_i is a training sample, y_i is the corresponding ground-truth label, and $M_\Theta(x_i)$ is the predicted label. Mathematically, the training procedure of the benign model is to minimize the loss function: $loss = \mathcal{L}(M_\Theta(x_i), y_i)$. HT detection is a binary classification task, and therefore, y_i is either 0 or 1. In this case, $\mathcal{L}$ is selected as the cross entropy. In addition, L_2 regularization and *dropout* strategies are also applied in our framework to avoid overfitting problem.

3.3 Backdoor Training

Based on the discussion in Sect. 3.1, backdoor training aims at building a mapping function that always gives zero value unless specific requirements are satisfied. Intuitively, a value checking logic plus a lock should suffice. Unfortunately, this naive approach needs hard-coding of constraints, which has no flexibility. In addition, this approach is very easy to detect due to its unique structure.

Our proposed backdoor training works in a totally different way. First, we select circuit samples and record their initial values of backdoor features. Next, we randomly apply various modification patterns multiple times to mutate the backdoor feature values, as shown in Fig. 4. Note that all mutation patterns are logically invariant. Meanwhile, changes applied in our work are controlled within a scale of <25 gates to satisfy the negligibility requirement.

After mutation, the modified patterns of backdoor features are considered as backdoor "signatures" to indicate whether it has been retrofitted by adversary or not. Then the task of backdoor training is to feed these malicious samples into the ML model to enforce it to remember these "signatures." In this case, the payload model works as a binary classifier, aiming at predicting whether input samples are with "signatures." This approach fulfills the desired constraint checking functionality as shown in Fig. 3.

Designing the structure of the payload model is even more challenging than the normal model. While a simpler structure is easier to train and harder to detect due to its small overhead, it often provides lower attack success rate for its limited capability. On the other hand, a complicated structure usually guarantees

Mutation Patterns	Affects
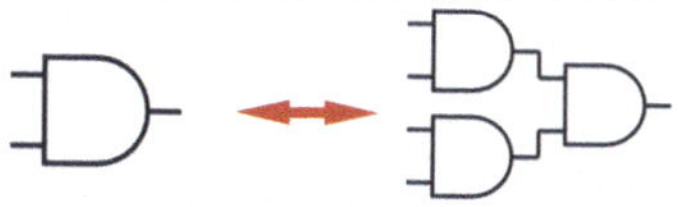	level ± 1 pout ± 1
	level ± 2 pout ± 2
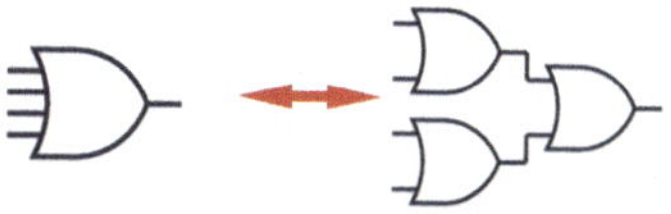	level ± 1 pout ± 1
	level ± 2 pout ± 2

Fig. 4 Example mutation patterns [13] used in our proposed work

the performance in terms of backdoor attack but comes at the cost of higher training cost as well as higher risk of being detected.

3.4 Trojan Injection

After backdoor training, we obtained the desired payload model. To complete the attack, we need to inject this payload model into the normal model. As described in Fig. 3, the desired functionality of payload model is to produce some perturbation that suffices to switch classifier prediction when the trigger condition is satisfied and maintain silence otherwise. The output of payload model can be designed as

$$output = -\lambda \cdot H(\bar{M}_{\bar{\Theta}}(x_i)) \cdot \mathcal{L}(M_{\Theta}(x_i), y_i)$$

where λ is the regularizer, $\bar{M}$ is the payload model, M is the normal model, and H is the Heaviside step function (unit-step function). In this case, when input circuit is recognized as "1" (with backdoor signature), $H(\bar{M}_{\bar{\Theta}}(x_i)) = 1$, and the output is a scaled inverse of normal model output. In terms of "0" label (without backdoor signature), $H(\bar{M}_{\bar{\Theta}}(x_i)) = 0$ and the output is 0. By combining the output layers, the normal model and payload model are assembled together. After pruning and nodes merging, the result is the desired backdoored ML model. The payload model is embedded into the normal model, and it hides behind the entire structure.

4   Defenses Against AI Trojans

In this section, we investigate possible mechanisms to alleviate the AI Trojan attacks for HT detection. There are three categories of defenses (trigger elimination, backdoor elimination, and backdoor mitigation). We explore five state-of-the-art defense strategies (Pruning, Bayesian Neural Networks, Neural Cleanse, Artificial Brain Simulation, and STRIP) selected from the three categories as shown in Fig. 5. The remainder of this section describes each of the defense strategies in detail.

4.1   Pruning

Pruning is the process of removing over-weighted connections in a network, aiming at accelerating the processing speed within the network, while reducing the size of the ML model. The principle behind pruning is the assumption that DNNs are commonly over-parameterized. The depth and huge amount of neurons grant the model with incredible ability to simulate the input–output relationship inherited in various tasks. However, the overwhelming size and complicated structure inevitably contain unused parameters. Therefore, pruning a network can be thought of as removing unused parameters from the over-parameterized network.

Pruning can be done by removing either redundant weights or nodes, as shown in Fig. 6. The pruning process is achieved by setting individual parameters to zero, which usually leads to sparse matrices in the network and can be further accelerated by sparse coding [14]. Notice that pruning is generally a model compression technique, not a specific strategy to defend against AI Trojan attacks. The key factor for pruning to be effective against AI Trojan is the removal of redundancy. It is a fact that the Trojan-embedded model works as benign unless attacker-chosen inputs are given, which means the backdoor trigger is a redundant component when dealing with clean inputs. Therefore, a feasible idea is to test the suspicious model with a

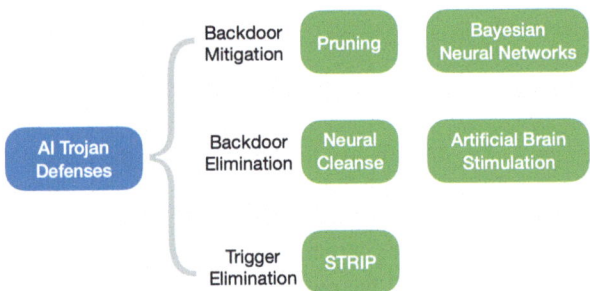

Fig. 5 Various defense strategies against AI Trojans evaluated in this chapter

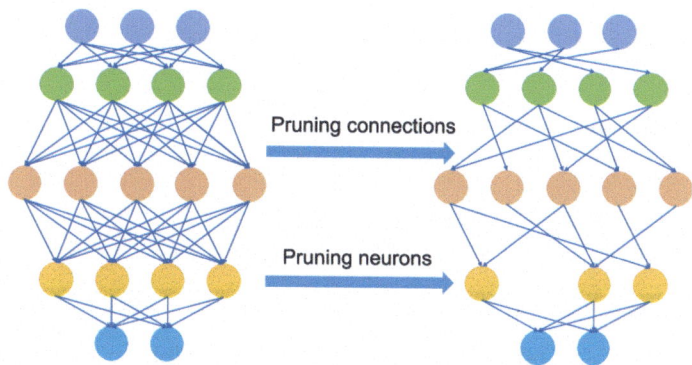

Fig. 6 Pruning can either remove redundant weights or nodes from the network

sufficient number of clean inputs, sort the nodes/weights by importance, and remove the unimportant parts.

Pruning is a feasible approach to defend against AI Trojan. Recent works have shown promising results by using pruning against backdoor attacks [10]. However, it has two major drawbacks: (1) Pruning strategy is fragile when the AI Trojan attack applies nodes/weights merging steps. (c). Intuitively, the redundancy has been evenly distributed through the merging of nodes. Therefore, there is no way to effectively find less-important nodes from the network. (2) When the model undergoes pruning, it can lead to loss of accuracy. Therefore, the users also need to consider the recovery of accuracy while pruning.

4.2 Bayesian Neural Networks

Given the drawbacks of pruning strategy, Bayesian neural network (BNN) is another option to defend against AI Trojan attacks. BNNs are a special type of DNNs as shown in Fig. 7. In traditional DNNs, weight values are real values and are commonly fixed after training. There is a fundamental difference between DNNs and BNNs. BNNs handle ML tasks from a stochastic perspective where all weight values are *probability distribution*, while DNNs use numerical weight values and utilize activation functions. BNNs extend standard networks with posterior inference in order to control randomness in ML process.

If BNN is used as the target ML model, both data poisoning and model injection methods can be blocked for the following reasons: (1) BNNs have natural resistance against data poisoning. BNNs produce output values with uncertainty, which severely limits the performance of any targeted attack. Also, in data poisoning attack, the goal is to train a model where a small change in input (noise) can cause significant change of output, which is protected by BNNs' regularization properties. (2) Model injection also suffers from the uncertainty possessed by BNNs.

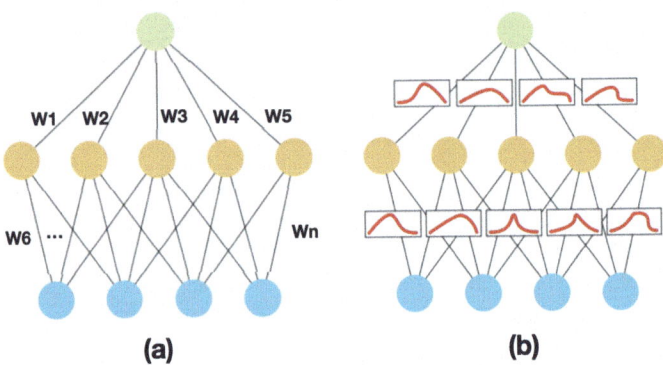

Fig. 7 Comparison of deep neural network (DNN) and Bayesian neural network (BNN). (**a**) DNN has weight values and utilizes activation functions to introduce non-linearity. (**b**) BNN is a stochastic neural network with probability distributions over the weights

The pruning technique is fragile against model injection attack due to the fact that the backdoor trigger nodes are merged with those benign ones. Without merging the two networks, the user can easily detect/remove the backdoor by pruning. In fact, it can be easy to identify the model structure since in most cases of MLaaS, the users typically specify the architecture of the expected ML model.

In traditional DNNs, edges connecting nodes contain only fixed weight values; therefore, merging two neural networks is straightforward. However, in case of BNN, there is no naive way to merge two probability distributions with different variables. In this case, even the joint distribution is not equivalent to the "add" operation for distributions. As a result, model injection attack may not be effective in BNNs due to the inability of merging nodes.

There are still limitations on applying BNNs against AI Trojan attacks. First, it is only feasible for partially outsourced scenario. In fully outsourced scenario, the ML model is provided to users as a black box. Even if the user specifies the structure and type of model, the adversary can somehow introduce randomness to the malicious model and pretend it to be a BNN. Second, BNNs themselves suffer from two main drawbacks. (1) BNNs are computationally expensive. Therefore, they are not suitable for large-scale problems in terms of efficiency. (2) BNNs enforce random variables to be in a cause–effect relationship. As a result, it is not suitable for hardware Trojan detection.

4.3 Neural Cleanse

Both pruning and applying BNNs are mitigation techniques to render the backdoor ineffective. In [19], the authors proposed a novel approach to directly detect the existence of backdoor in the ML model. The key intuition of the detection method

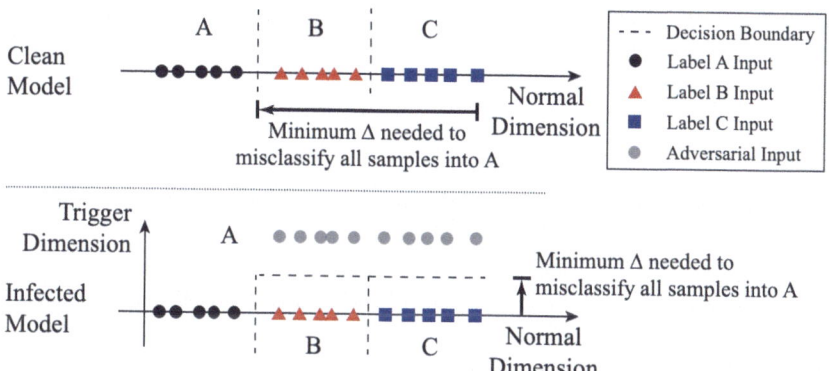

Fig. 8 A simplified illustration of the neural cleanse

is that, for an infected model, it uses small modification to cause misclassification of the target label, which is shown in Fig. 8. The top figure shows a clean model, where significant modification is needed to shift the samples of type B and type C across the decision boundaries to be misclassified into label A. The bottom figure shows the infected model, where it needs minor modification to move samples across the boundary.

This intuition is based on the assumption that a small change of input (noise) can cause significant change of output due to existence of a backdoor. It works by crafting perturbed input samples to test if significant change happens at the output. Unlike pruning or BNN, neural cleanse is not interested in removing the backdoor trigger or prevent attack from happening, it focuses on maximizing the chances of raising red flag when backdoor exists.

To achieve this, the authors in [19] iterate through all labels of the model and determine if any label requires significantly a smaller amount of modification to achieve misclassification. Application of neural cleanse for hardware Trojan detection is easier since there are only two labels, clean or HT implanted. Therefore, the working steps can be simplified as: (i) for a given design, continuously add "noise" by randomly applying adversarial patterns, (ii) check if any of the modification unintentionally activated the trigger so that there is significant change at the output.

Neural cleanse is a promising technique for mitigating backdoor attacks. However, it has two major limitations: uncertainty and expensive. (1) There are no precise guidelines for making these modifications to maximize the opportunity to trigger the backdoor. This is analogous to random test generation for hardware Trojan detection, where the chances of triggering Trojans are very low, and in the worst case, the backdoor is out of the coverage of generated test patterns. (2) Since neural cleanse relies on somewhat random choices, it can be expensive in terms of effort to apply this strategy. One major motivation for applying ML to detect HT is the relatively low cost compared to traditional test generation schemes. If we decide to use neural cleanse, the users have to pay significant effort to defend against possible attacks toward the ML models themselves. In other words, applying ML for HT detection is not beneficial if we have to rely on time-consuming neural cleanse for defense purposes.

4.4 Artificial Brain Stimulation (ABS)

Artificial brain stimulation (ABS) [12] is a novel approach that scans the entire ML model seeking potential backdoor by analyzing inner neuron behaviors through a stimulation method. ABS is performed in 3 steps. First, it locates suspicious neurons inside the ML model, which substantially elevates the activation of a particular output label regardless of the provided input. Neurons acting like this are considered as potential backdoor. Second, with the potential nodes obtained, the authors in [12] craft the Trojan triggers through reverse engineering, implemented as an optimization procedure. The crafted triggers are attached to clean inputs. Finally, this artificial Trojan-embedded input is given to the model to confirm whether it produces incorrect labels. If yes, the backdoor is detected.

ABS is similar to neural cleanse. However, they have one major different—unlike randomly generating inputs that occasionally activates the trigger in neural cleanse, ABS starts by locating suspicious targets. If we use the analogy of hardware Trojan detection, neural cleanse acts like random test generation, while ABS works like constrained-random or directed test generation that aims at activating potential backdoor triggers (e.g., rare signals).

Specifically, the authors in [12] evaluate ABS on 177 Trojan-embedded models by various attack methods with various trigger sizes and shapes, where promising defense performances are obtained. However, ABS still assumes that a user has full access to the ML model, which is not realistic in fully outsourced scenario. Moreover, ABS is designed for solving targeted attacks. In case of untargeted attacks, ABS is likely to be less effective.

4.5 STRIP

STRIP is a strong defense against Trojan Attacks on DNNs [4]. STRIP is an experience-based trigger-detection method that aims at detecting suspicious backdoor triggers from inputs. Through heuristic analysis as well as experimental evaluation, the authors have observed a drastic difference in entropy between clean inputs and Trojan-inserted inputs. It works by fusing the input samples with multiple clean samples. Then STRIP applies the fused input to the backdoored model and calculates the entropy of model outputs. In general, a low entropy in predicted classes usually violates the input-dependence property of a benign model and implies the presence of a malicious Trojan-embedded input.

There are still many limitations for STRIP, especially when applied for HT detection. (1) STRIP was proposed and examined in computer vision domain. STRIP is assumption-based, but whether the entropy assumption still holds for hardware circuit features remains unclear. (2) STRIP utilizes the distributed output from the last layer of the model, which is not applicable in fully outsourced scenario, where users only receive the prediction results from the model. (3) STRIP assumes that the user has access to Trojan-embedded samples under the threat model, which is usually not true. Therefore, STRIP is not effective in realistic scenarios.

5 Experiments

5.1 Experimental Setup

The experimental evaluation is performed on a host machine with Intel i7 3.70 GHz CPU, 32 GB RAM, and RTX 2080 256-bit GPU. We developed code using Python for model training. We used PyTorch as the machine learning library. To enable comprehensive evaluation, we deploy the experiments utilizing 50 gate-level netlist benchmarks from Trust-Hub [1]. Features are extracted from benchmarks and formatted into PyTorch tensors, making them compatible with any ML models requiring tensor inputs. The structure for normal model is described in Sect. 3.2. Based on Sect. 3.3, we apply the following models when designing **our payload model**:

- **MLP:** A multiple layer perceptron (MLP), composed of 3 fully connected layers
- **Lenet:** A Lenet-5 [9] like structure, composed of 3 convolution layers followed by 2 fully connected layers
- **GoogleNet:** A GoogleNet [17] like structure, with a depth of 22 layers

We report baseline accuracy, attack success rate (ASR), as well as backdoor accuracy evaluate the performance. To evaluate the effectiveness of our approach, we compare with the following **state-of-the-art** attack and defense:

- **GAE:** State-of-the-art adversarial attack based on generating adversarial examples [13]
- **STRIP:** State-of-the-art defense against AI Trojan attacks [4]

5.2 Comparison of Attack Performance

Figure 9 compares the performance of three different implementations. In each figure, baseline accuracy, backdoor accuracy, and attack success rate are provided. The x-axis represents the upper bound on the number of mutations applied during

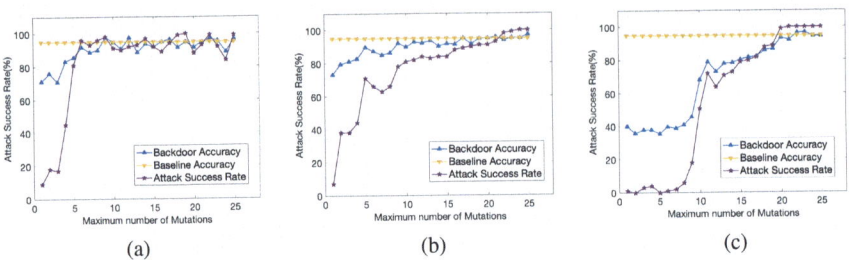

Fig. 9 The attack success rate of our framework using three different payload models under different thresholds on a number of mutations. (**a**) MLP. (**b**) Lenet. (**c**) GoogleNet

backdoor training, where larger x-value represents more modifications to the input samples. In our experiment, the normal model achieves 98.5% accuracy for normal samples. All three models' backdoor accuracies are slightly lower than the baseline accuracies. This difference comes from the effect of payload model. This is supported by the observation that the backdoor accuracy is nearly proportional to the ASR. The closer ASR is to perfection, the closer backdoor accuracy is to the baseline. In other words, it represents the performance of backdoored model "mimicking" the normal model's behavior. In terms of attack success rate, the simpler (lightweight) payload model implies faster convergence to perfection. For example, MLP needs about 5 mutations, while GoogleNet requires 20 mutations to reach 100% ASR. However, as we can see, the ASR of MLP is unstable. Even after it hits perfection, it oscillates at a 10% amplitude. Instead, complicated model like GoogleNet requires more modifications to reach convergence, but it becomes very stable once reaches 100% success rate. This is expected due to simple models' limited capability in handling complex features. A larger number of mutations bring expanded feature space, and it is likely for these lightweight models to get overfitted. In other words, some normal samples may satisfy the payload model and get their classification result switched. Therefore, we need to carefully select the mutation number for simple structures.

Figure 10 compares the ASR of our proposed method with state-of-the-art attack, GAE [13]. As we can see, GAE's ASR is much lower than the proposed method. This huge difference comes from the design strategy. GAE applies mutations on circuits and then directly feeds them into models to alter its outputs. In this chapter, we extract backdoor features and feed them into an extra model. Intuitively, this extra model acts as both an extractor and an amplifier. It recognizes backdoor features and enables fusion of its output with results from the normal model. As a result, a small amount of mutations suffices to alter the classification result. In contrast, GAE does not have such amplifier, and it usually requires a large number of mutations to create changes in the output layer. Therefore, it provides inferior attack performance. GAE also faces the risk of being detected due to a larger number of mutations.

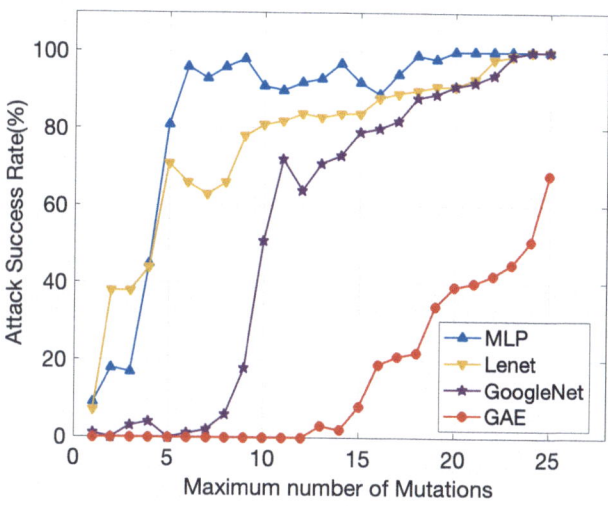

Fig. 10 The attack success rate comparison between proposed algorithm and the state-of-the-art adversarial attack with <25 mutations

Table 2 Comparison of Training Cost and Data Resources

Models	Time(s)	Epochs	Division Malicious/Benign	# Mutation
MLP	0.6	50	2/8	6
Lenet	1.7	200	2/8	18
GoogleNet	72.4	500	5/5	21
GAE [13]	1.0	200	4/6	44

5.3 Overhead Analysis

Table 2 compares the training cost and data resources of various methods. The first three rows represent our approach.

The MLP approach is the most economic in terms of training cost. It can be trained within 50 epochs with each epoch taking 0.6 s and only requires 20% of the training samples to be malicious. However, GoogleNet is very costly, and it needs 500 of 0.37 s training epochs. GAE requires moderate training cost, comparable to Lenet. However, it requires a large number of mutations and still provides inferior attack performance compared to our proposed method.

5.4 Robustness Against STRIP-Based Defense

We further evaluate the proposed attack's robustness against the state-of-the-art defense scheme, STRIP [4]. STRIP aims at identifying if a given input is clean

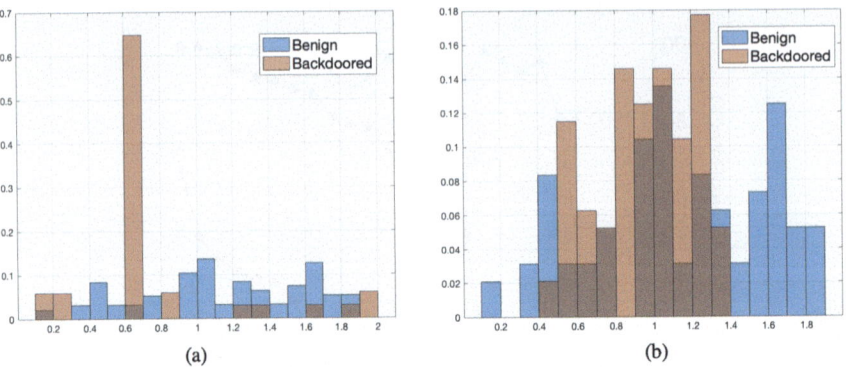

Fig. 11 Entropy distribution of clean and backdoored inputs. (**a**) Backdoored inputs generated by GAE can be easily detected. (**b**) Backdoored inputs generated by our method's entropy are hard to distinguish from benign inputs. (**a**) GAE against STRIP. (**b**) Proposed against STRIP

or contains a backdoor trigger. It works by fusing the input sample with multiple clean samples. Then STRIP applies the fused input to the backdoored model and calculates the entropy of model outputs. This defense strategy relies on the observation that backdoored inputs tend to produce lower entropy outputs compared to the clean ones, so that by checking their entropy distributions, backdoored inputs can be clearly distinguished.

Figure 11 shows the entropy of outputs from GAE and our proposed method for both clean and backdoored inputs. As we can see, the distribution of entropy for backdoored data overlaps with the distributions of entropy of the clean data for our approach. However, GAE's entropy can be clearly distinguished from the normal one. We consider the following two important reasons for this scenario: (1) We intentionally select backdoor features that are orthogonal to normal features. Therefore, applied mutations do not affect the normal features, which avoids drastic changes in output entropy. (2) The mutations in GAE are gradient-driven, where feature values are erected to the gradient direction, leading to a small entropy. Our proposed method is able to bypass the state-of-the-art defense (STRIP), while state-of-the-art attack (GAE) fails.

6 Summary

While machine learning (ML) techniques are widely applied in hardware Trojan (HT) detection, ML algorithms are vulnerable toward Trojan attacks. In this chapter, we exploit this fundamental vulnerability to propose a backdoor attack scheme. We present an efficient mechanism to design and inject AI Trojans into ML models for HT detection. The infected model can hide in plain sight since it can provide expected classification for regular inputs. However, it will produce misclassification for specific attacker-chosen inputs. Extensive experimental evaluation using three implementation models demonstrated that our approach can achieve 100% attack success rate with very few modifications compared to state-of-the-art adversarial attack for ML-based HT detection. We also discuss potential countermeasures to defend against AI Trojan attacks.

References

1. TrustHub.org: Trust-HUB,. http://trust-hub.org/benchmarks/trojan.
2. Daniel Arp et al. Effective and efficient malware detection at the end host. In *18th USENIX*, Montreal, Quebec, 2009.
3. George E. Dahl, Jack W. Stokes, Li Deng, and Dong Yu. Large-scale malware classification using random projections and neural networks. In *ICASSP*, pages 3422–3426, 2013.
4. Yansong Gao et al. STRIP: A defence against Trojan attacks on deep neural networks. In *ACSAC*, pages 113–125, 2019.
5. Kathrin Grosse et al. Adversarial perturbations against deep neural networks for malware classification. *CoRR*, 2016.
6. Tianyu Gu et al. BadNets: Evaluating backdooring attacks on deep neural networks. *IEEE Access*, 7:47230–47244, 2019.
7. Kento Hasegawa et al. A hardware-Trojan classification method using machine learning at gate-level netlists based on Trojan features. *IEICE TFECCS*, 100(7):1427–1438, 2017.
8. Kento Hasegawa et al. Trojan-feature extraction at gate-level netlists and its application to hardware-Trojan detection using random forest classifier. In *ISCAS*, pages 1–4. IEEE, 2017.
9. Yann LeCun et al. Lenet-5, convolutional neural networks. http://yann.lecun.com/exdb/lenet, 20(5):14, 2015.
10. Kang Liu et al. Fine-pruning: Defending against backdooring attacks on deep neural networks. In *ISRAID*, pages 273–294. Springer, 2018.
11. Yingqi Liu et al. Trojaning attack on neural networks. 2017.
12. Yingqi Liu, Wen-Chuan Lee, Guanhong Tao, Shiqing Ma, Yousra Aafer, and Xiangyu Zhang. ABS: Scanning neural networks for back-doors by artificial brain stimulation. In *SIGSAC*, pages 1265–1282, 2019.
13. Nozawa et al. Generating adversarial examples for hardware-Trojan detection at gate-level netlists. *JIP*, 29:236–246, 2021.
14. Bruno A Olshausen and David J Field. Sparse coding of sensory inputs. *Current Opinion in Neurobiology*, 14(4):481–487, 2004.
15. Zhixin Pan and Prabhat Mishra. Automated detection of Spectre and Meltdown attacks using explainable machine learning. In *2021 IEEE International Symposium on Hardware Oriented Security and Trust (HOST)*, pages 24–34. IEEE, 2021.
16. Joshua Saxe and Konstantin Berlin. Deep neural network based malware detection using two dimensional binary program features. In *10th MALCON*, pages 11–20, 2015.
17. Christian Szegedy et al. Going deeper with convolutions. In *CVPR*, 2015.
18. Bolun Wang et al. Neural cleanse: Identifying and mitigating backdoor attacks in neural networks. *SP*, 530546, 2019.
19. Bolun Wang, Yuanshun Yao, Shawn Shan, Huiying Li, Bimal Viswanath, Haitao Zheng, and Ben Y Zhao. Neural cleanse: Identifying and mitigating backdoor attacks in neural networks. In *2019 IEEE Symposium on Security and Privacy (SP)*, pages 707–723. IEEE, 2019.
20. Qinglong Wang et al. Adversary resistant deep neural networks with an application to malware detection. In *Proceedings of the 23rd ACM SIGKDD*, pages 1145–1153, 2017.
21. Hasini Witharana and Prabhat Mishra. Speculative load forwarding attack on modern processors. In *IEEE/ACM International Conference on Computer-Aided Design*, pages 1–9, 2022.

Part V
Acceleration of Explainable AI

Hardware Acceleration of Explainable AI

1 Introduction

Due to the widespread adaptation of artificial intelligence (AI) and machine learning (ML) algorithms, numerous methods and techniques have been invented to optimize their performance [3, 8, 9, 14, 19, 21, 22]. Hardware-based acceleration is a widely used approach to enhance the model performance by optimizing the hardware architecture. These hardware accelerators are specially designed to efficiently execute ML computations, such as matrix multiplication. For example, inference hardware accelerators are implemented on the edge devices while supporting the forward pass phase. Similarly, training hardware accelerators are designed by optimizing the hardware, primarily to support the calculation in the weights update phase during the training process.

The "black-box" nature of ML models raises serious concerns about their interpretability and trustworthiness. It is challenging for users to understand the underlying reasons for specific decisions or predictions, leading to questions and doubts regarding the model's output. Explainable AI (XAI) addresses this challenge to provide insights into the ML model's decision-making process [14, 16, 18, 23]. Due to the inherent inefficiency in XAI algorithms, they are not applicable on various domains, including real-time and safety-critical systems [1, 17]. These algorithms treat the explanation process as an extra procedure and perform the interpretation outside the learning model, which makes them inefficient in practice. Specifically, it solves a complex optimization problem that consists of numerous iterations of time-consuming computations. In other words, such a time-consuming interpretation is not suitable for time-sensitive applications with soft or hard deadlines. In soft real-time systems, such as multimedia and gaming devices, inefficient interpretation can lead to unacceptable degradation of Quality-of-Service (QoS). In hard real-time systems, such as safety-critical systems, missing task deadlines can lead to catastrophic consequences.

There are recent efforts to achieve fast XAI utilizing various hardware accelerators, including Field Programmable Gate Arrays (FPGA), Graphic Processing Units (GPU), and Tensor Processing Units (TPU) [15]. Specifically, they can transform XAI procedures into computation of matrix operations and exploit the natural ability of hardware accelerators for fast and efficient parallel computation of matrix operations. FPGAs are flexible and reconfigurable devices that enable efficient and parallel processing of complex computations. This characteristic makes them suitable for implementing hardware accelerators. GPU consists of a large number of cores and high-speed memory for performing efficient matrix computation and parallel computing. Similarly, TPU is an Application-Specific Integrated Circuit (ASIC) developed specifically to accelerate the computations in deep neural networks [7, 10, 11], with extremely high throughput and fast performance with low-memory footprint. It is a compatible choice to deploy in our proposed framework.

In this chapter, we will discuss the hardware accelerator implementation of XAI techniques using FPGA as well as GPU architectures. Chapter "Explainable AI Acceleration Using Tensor Processing Units" will describe hardware acceleration of XAI methods using Tensor Processing Units. We first provide a brief introduction to FPGA and GPU. Next, we describe hardware accelerators for different explainable AI methods.

1.1 Graphics Processing Unit

Graphics Processing Unit (GPU) is the widely adopted hardware accelerator for machine learning algorithms. It uses a large number of computing units and long pipelines. There are two primary reasons for the superior performance of GPUs compared to software (CPU)-based acceleration: (i) multi-threading utilizes multi-core parallel computing of neural network data and (ii) higher memory access speed and floating-point computing capabilities. Figure 1 shows an example of NVIDIA GPU architecture with a large number of processor cores organized in multiple clusters.

NVIDIA GPUs consist of Single-Instruction Multiple-Thread (SIMT) architecture, where thousands of parallel floating-point units are used to efficiently process problems with data-level parallelism. This makes GPUs well-suited for speeding up machine learning as well as scientific and multimedia applications with matrix multiplication. To achieve the desired speedup, there are several multithreaded SIMD processing units inside the GPU, which are known as Streaming Multiprocessors (SMs). A bundle of 32 threads is referred to as a warp. The out-of-order execution of threads in a warp on SMs enhances the processing performance. Because of the unique architecture of GPU, all the threads in a warp should execute the same instruction, and there should be an evenly distributed load between them. Compared to CPUs, GPUs have a limited number of registers per SMs, which can lead to limitations on intermediate values. Therefore, any calculation with divergent

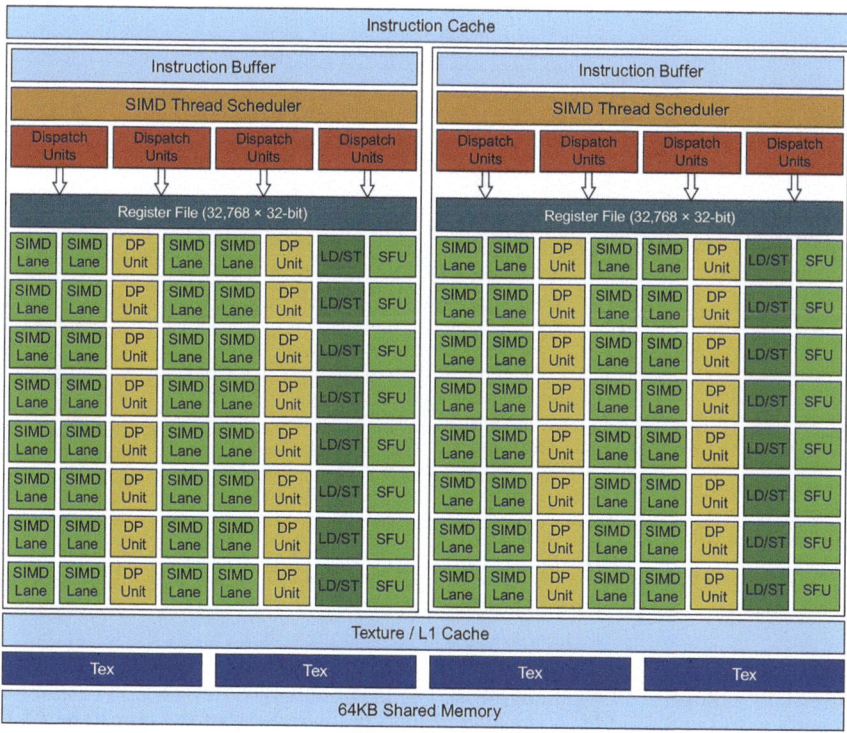

Fig. 1 An example NVIDIA GPU architecture with a large number of processor cores organized in multiple clusters. Image Credit [20]

branches per thread in a warp, unbalanced load per thread in a warp, or excessive register usage can result in a reduction in GPU performance.

1.2 Field Programmable Gate Array

Field Programmable Gate Array (FPGA) is one of the most widely used hardware for accelerating machine learning algorithms due to its reconfigurability and capability for parallelism. Besides its application as a hardware accelerator, FPGA can be found in various scientific instruments for signal processing, satellites, high-performance network routers, etc. Shorter design time, lower cost in low volumes, and greater flexibility compared to Application-Specific Integrated Circuits (ASICs) are some reasons behind this extensive range of applications. We can visualize FPGA as a blank canvas that can be used to design and implement our own computing cores or circuits.

The basic FPGA architecture has three main components: configurable logic blocks (CLBs), programmable interconnects, and input/output (I/O) blocks. As

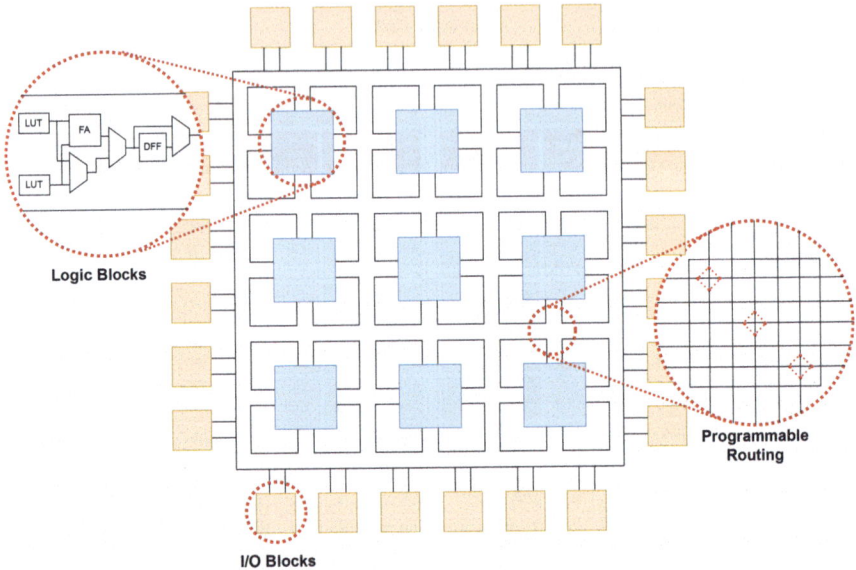

Fig. 2 An overview of a typical FPGA architecture

shown in Fig. 2, I/O blocks serve as the programmable interface between the outside world and the FPGA, located around the periphery of the chip. The CLB is the fundamental element in an FPGA, consisting of a Look-Up Table (LUT), combinational logic, and a storage element, as shown in Fig. 2. The programmable interconnects, also known as fabric, provide connectivity between CLBs as well as CLBs and I/Os.

The remainder of this chapter describes hardware accelerators for three different explainable AI methods. Specifically, Sect. 2 describes FPGA-based acceleration of heatmap visualization. Next, Sect. 3 describes FPGA-based acceleration of saliency map. Finally, Sect. 4 describes GPU-based acceleration of Shapley value analysis.

2 FPGA-Based Acceleration of Heatmap Visualization

The discussion in this section is based on the work of Bhat et al. [5]. Visualization is a critical component in explainable AI as it provides post-prediction interpretability. Specifically, it takes a trained ML model as input and provides feature importance scores. These importance scores provide quantified measurements of the importance of the input features to the model output. These importance scores, also known as feature attribution scores, are particularly useful in the field of image classification, especially when applied to Convolutional Neural Networks (CNNs). Heatmap visualization is a commonly employed technique to represent these scores and gain insights into CNN's classification process. The heatmap visualization of the input pixels highlights the pixels of an image that have the highest contribution to the model's output, providing insight into the features that drive the model's output.

Model training and inference are the two major steps in the development and implementation of a machine learning model. The training step comprises three sub-phases: forward pass (FP), backpropagation (BP), and weights update (WU). In the FP phase, input data are propagated through the network from the input layer to the output layer, producing the network's output. After the FP, BP is used to propagate the output error from the output layer back to the network's input layer. Then in the WU phase, gradients are calculated with respect to each model parameter, which introduces computational overhead to the phase. Storing all the intermediate activations during FP is required to calculate those gradients, which adds memory overhead to the WU phase. These two overheads contribute to making the WU step the most expensive step in training. The inference step only has the FP phase, where new (unseen) data are taken as input to the trained model to produce the desired output.

Regarding the dataflow of feature attribution algorithms, they typically consist of two common steps: FP and BP. During the BP step of feature attribution, relevance scores for input features are calculated by backpropagating through the model. Even though WU is not a step in the feature attribution, it calculates the local activation gradient layer by layer in the BP step.

To repurpose the existing AI accelerators to support visualization-based XAI, additional resources need to be added to the inference accelerators to support BP. While training accelerators can inherently support vanilla gradient-based feature attribution, they may require modifications to accommodate alternative visualization algorithms. Moreover, there are only a few hardware accelerators specially designed for XAI. Therefore, we focus on the development of a flexible accelerator for three feature attribution algorithms that enable real-time XAI on edge devices [5]. As mentioned earlier, inference hardware accelerators are usually implemented on edge devices. The BP step needs to be added to enable the implementation of an XAI accelerator on edge devices. Additionally, to support flexibility for multiple feature attribution algorithms in the same accelerator, dataflows of feature attribution algorithms are analyzed to determine their hardware/software overhead compared to inference.

This section analyzes three gradient backpropagation-based feature attribution methods, namely saliency map, DeconvNet, and Guided Backpropagation [5]. When it comes to handling gradient signals in the presence of a ReLU activation layer within a Deep Neural Network (DNN), the above methods differ in their approaches. The ReLU activation layer outputs the layer input directly if it is positive; otherwise, it outputs zero, and its functionality is described in Eq. 1 when the ReLU activation is at layer L. In this context, f_i^L represents the output of layer L, which is the input to the ReLU activation layer. Similarly, f_i^{L+1} represents the output of the ReLU activation layer, which then becomes the input to the next layer $(L + 1)$.

$$f_i^{L+1} = \text{ReLU}\left(f_i^L\right) = \max\left(f_i^L, 0\right) \tag{1}$$

The first gradient backpropagation-based feature attribution method that we will discuss is the saliency map method. The saliency map is used to visually depict the importance of input features in the decision-making process. These importance scores $(R_i(x))$ are calculated for each input feature (x_i) based on the partial derivative of the model's output for specific class c $(f_c(x))$ with respect to that feature. The calculation is described by Eq. 2.

$$R_i(x) = \frac{\partial f_c(x)}{\partial x_i} \tag{2}$$

During BP process at the ReLU activation layer, the gradient signals for neurons that had negative activation values during FP are set to zero. Equation 3 depicts the above calculation, where R_i^L is the relevance score of layer L, f_i^L is the output of layer L during FP, and f^{out} is the model output during FP. As shown in Eq. 3, to support BP for ReLU activation, it is required to store the indices of the negative activation values.

$$R_i^L = \left(f_i^L > 0\right) \odot R_i^{L+1}, \quad \text{where } R_i^L = \frac{\partial f^{out}}{\partial f_i^L} \tag{3}$$

The second feature attribution method based on gradient backpropagation that we will discuss is the DeconvNet method. The original purpose of deconvolution was to recreate the CNN inputs from its outputs in an unsupervised manner. However, due to its ability to visualize the most important features, deconvolution has also been used as an XAI technique. During BP, DeconvNet comprises deconvolution layers, which replace the convolution layers, and unpooling layers, which replace the max-pooling layers. These four layers are further explained in the hardware acceleration implementation part. Moreover, these replacements justify the fact that deconvolution can be considered as an inverse operation of CNN during FP.

$$R_i^L = \left(R_i^{L+1} > 0\right) \odot R_i^{L+1} \tag{4}$$

As depicted in Eq. 4, unlike the saliency map method, the DeconvNet method directly outputs the gradient value if it is positive. If the gradient value is negative, it is set to zero. This implies that features with a positive effect on the network's output are considered more important. So, Eq. 4 relies solely on the values calculated during the BP phase, resulting in the smallest memory overhead.

The third feature attribution method is guided backpropagation. It has fundamental backpropagation properties of both the saliency map method and DeconvNet. Equation 5 describes the functionality of Guided Backpropagation, which sets the gradient signal to zero in two cases: (i) when the neuron has a negative activation value during FP, similar to the saliency map method, and (ii) when the gradient value itself is negative, similar to DeconvNet. Due to its similar behavior to the saliency maps, it also requires storing necessary indices, which introduces a memory overhead.

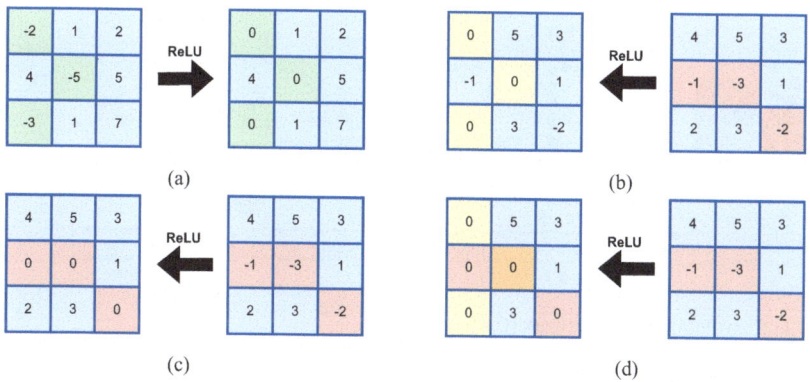

Fig. 3 Comparison of feature attribution method at ReLU activation layer. (**a**) Forward pass. (**b**) Saliency map. (**c**) DeconvNet. (**d**) Guided backpropagation

$$R_i^L = \left(f_i^L > 0 \right) \odot \left(R_i^{L+1} > 0 \right) \odot R_i^{L+1} \tag{5}$$

Figure 3 illustrates the functionality at the ReLU activation layer during the FP and the BP using the three gradient backpropagation-based feature attribution methods mentioned above. With an initial understanding of gradient backpropagation-based feature attribution methods, let us delve into their hardware implementation on edge devices. This implementation must be flexible enough to support all three feature attribution algorithms, and specially it should fit into the limited resources environment of the edge device. Therefore, the implementation should be optimized to utilize available resources while effectively accelerating the feature attribution algorithms. The tile-based design is the perfect solution for this scenario, as it allows us to divide the entire process into subprocesses that fit with device constraints. The tile-based architecture naturally supports parallelism, and this maximizes the inherent parallelism in the feature attribution algorithms. This tile-based architecture is added on top of the layered CNN architecture.

As shown in Fig. 4, the CNN consists of four main layers: convolution layers, pooling layers, fully connected (FC) layer, and activation layers. The convolution layer serves as the input layer and major element in the CNN. This layer is used to create the feature maps, which are the abstract representations of the input image obtained by convolving the image with a set of filters. These filters, also known as convolution kernels, extract information from the input image. These feature maps also referred to as activation maps are then fed into the non-linear activation function.

The pooling layers downsample the feature maps, gradually reducing their spatial dimensions while saving the important characteristics. There are two commonly used pooling methods: max-pooling and average-pooling. Max-pooling selects the maximum value within each sub-region of a feature map and effectively disregards others. Average-pooling performs the same functionality as max-pooling with one

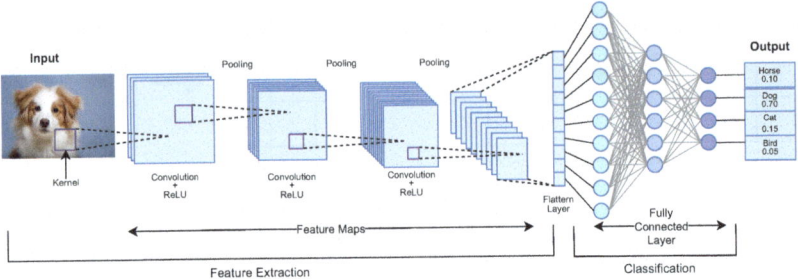

Fig. 4 Convolution neural network (CNN) layers

exception. Instead of selecting the maximum value, average-pooling calculates the average value of the sub-region and forwards it to the next layer. After having multiple iterations of convolution, activation and pooling, the final output is flattened into a vector. This flattened vector is then passed through the FC layer. The final layer of the CNN is the FC layer, which classifies the input image into several classes with probabilities.

In the proposed hardware implementation by Bhat et al. [5], the input image for the CNN, model parameters, and tiles are stored in the Dynamic Random Access Memory (DRAM) on FPGA. As explained earlier, most feature attribution algorithms have both FP and BP phases. In this context, CNN belongs to the FP phase. Thus, the convolution layer, pooling layer, non-linear layer, and FC layers have been implemented in the FPGA. The input and convolution kernel tiles are stored in on-chip buffers inside the convolution block. To create the feature maps using input, input values need to be multiplied with corresponding kernel values (weights) and then summed to obtain the final value. This operation is called multiplication and addition (MAC), and this operation is a very common mathematical operation in fields that interact with matrices. Thus, it is a common step in Digital Signal Processing (DSP). Therefore, the dedicated DSP units in the FPGA can utilize this operation to accelerate the overall performance. Parallelism plays a crucial role in hardware acceleration. To leverage parallelism, loop unrolling is a commonly employed technique. Loop unrolling involves duplicating loop iterations to perform multiple computations simultaneously. Usually, the unrolling factor is determined by the resource constraints, and here it is configurable at the design time.

The implemented hardware architecture consists of a ReLU unit as the non-linear activation block. Due to its functional simplicity, this block is implemented in the output buffer, allowing for in-place modification of the value. After the computation, the output tile is stored in the DRAM and used as input for the next layer. Within the store-back operation, the max-pooling is incorporated. Consequently, only the maximum values of each sub-region are stored back into the DRAM, reducing memory requirements and data movement. The final output of the convolution block is fed into the FC block, which is nothing but a vector–matrix multiplication (VMM). The authors have implemented a separate VMM block while portioning the input vector and weight matrix and loop unrolling to accelerate the computation.

Partitions of both the input vector and weight matrix are loaded into on-chip buffers, and the final output tile is stored back into the DRAM.

The BP phase initiates with this output value and progresses through the network's output to the input while computing the activation gradient. The gradient values, as explained in Eq. 2, are calculated using the chain rule of derivation. Due to the hardware constraints of the edge device, the BP phase is designed to reuse the functional blocks utilized in the FP phase. In this hardware architecture, FC is the first layer in the BP phase, corresponding to the VMM block. The gradient of a vector–matrix multiplication operation is a matrix–vector multiplication. As a result, the same VMM block can be reused, with the weights loaded into on-chip buffers allocated for inputs during the FP phase. Similarly, inputs that were output during the FP phase are loaded into on-chip buffers allocated for weights during the FP phase.

During the BP phase, the second layer encountered is the convolution layer. In this hardware architecture, the convolution layer is composed of sequential computations involving convolution block, ReLU layer, and max-pooling in the FP phase. Because of that, the gradient calculation of max-pooling, ReLU layer, and convolution block is performed in reverse order during the BP phase. The gradient of max-pooling is achieved by unpooling operation, which involves up-sampling the input values by placing back the maximum values selected in the pooling operation at its initial index, while other values in the sub-region are set to zero. This necessitates additional variable storage to store the index of the largest value of each sub-region during the pooling operation. These indexes are stored on-chip to minimize the data movement between DRAM and on-chip as a 2-bit value.

The gradient of ReLU activation is obtained by utilizing a 1-bit mask value stored in the on-chip Block RAM (BRAM). During the FP phase, this mask takes a value of one if the output of the ReLU is positive, and it takes a value of zero if the output is negative. This mask is then utilized during the BP phase to compute the gradient, as explained in Fig. 3.

Finally, in the convolution layer, the computation of gradients can be performed in a similar manner to the FP phase, with two key differences. During the FP phase, input feature maps are filtered to obtain output feature maps, while during the BP phase, this process is reversed. To achieve this, the same architecture of the convolution block is utilized in the BP phase by taking the transpose of the layer's weight parameters' channel dimensions and flipping the values of each filter by 180 degrees. Additionally, the inputs and output need to be interchanged, as explained in the BP phase computation in the VMM.

The proposed feature attribution accelerator is designed as High-Level Synthesis (HLS) library, which is used to prototype the design on a small FPGA. As explained earlier, this implementation can support feature attribution for CNN. Template functions have been implemented in the HLS library, which are necessary for different layers of the CNN during both the FP and BP phases. These template functions are selected based on the specific architecture of the CNN model being executed. Furthermore, the developed HLS library has the reconfigurable option to support several gradient backpropagation-based visualization methods. The experimental results of this work demonstrate the flexibility and reconfigurability of the HLS library and the utilization of FPGA resources such as BRAM (maximum 0.3% overhead), DSP (maximum 0.5% overhead), flip-flops (FF) (maximum 7.5% overhead), and look-up tables (LUT) (maximum 27% overhead) [5]. The implementation also shows efficient computation parallelism while having a latency cost (maximum 23.22 ms overhead) due to the additional BP phase with respect to inference [5].

3 FPGA-Based Acceleration of Saliency Map

The discussion in this section is based on the work of Asgari et al. [4]. An autonomous agent is a very popular concept in the modern world, and it is a computational system that makes decisions by itself to achieve some goals by observing and reacting to its surroundings. To acquire sensory data from their environment, these autonomous agents need advanced sensor systems, and an artificial vision system is one of those. Usually, these vision sensors send substantial volumes of visual data to the agent's real-time processing unit, posing a challenging and critical task for the processing unit. However, if the vision system is observing a steady environment, that means if there is no change in the environment, the agent will behave the same as before; nothing will change. Even though the agent behaves the same as before, it needs to process almost the same set of vision data in real time. However, if the vision system transmits data only when there is a change in the input, it can reduce many resources.

These types of vision sensors are called event-based vision sensors, which are suitable for low-latency and low-power vision processing applications. However, event-based vision systems still require powerful machine vision hardware and high-speed communication links to process and transmit these event-based sensor data. Especially the processing stage needs to be capable of parallel processing all sensory information. So, this still requires a considerable amount of resources. Part-by-part sequential processing is a good strategy for a resource-constrained environment. This method's performance should not affect the overall system performance. Now the question is, how can we divide the visual data into smaller sub-regions or what is the base for partitioning? We can answer this question by exploring the visual system of animals.

Because in an event-based vision sensor, their pixels produce events only in response to changes in the input, and if there is a way to identify and quantify that change and importance with respect to the vision, we can divide the visual field based on this value. A saliency map is a way of doing this. Now, we can choose the sub-region with the highest value by analyzing the saliency map and prioritize its processing. Then we can shift the focus of attention from one sub-region to another in descending order of saliency. This technique is widely known as "selective attention," which is used by the visual system of various animals to optimize vision processing with limited resources. It involves identifying the most

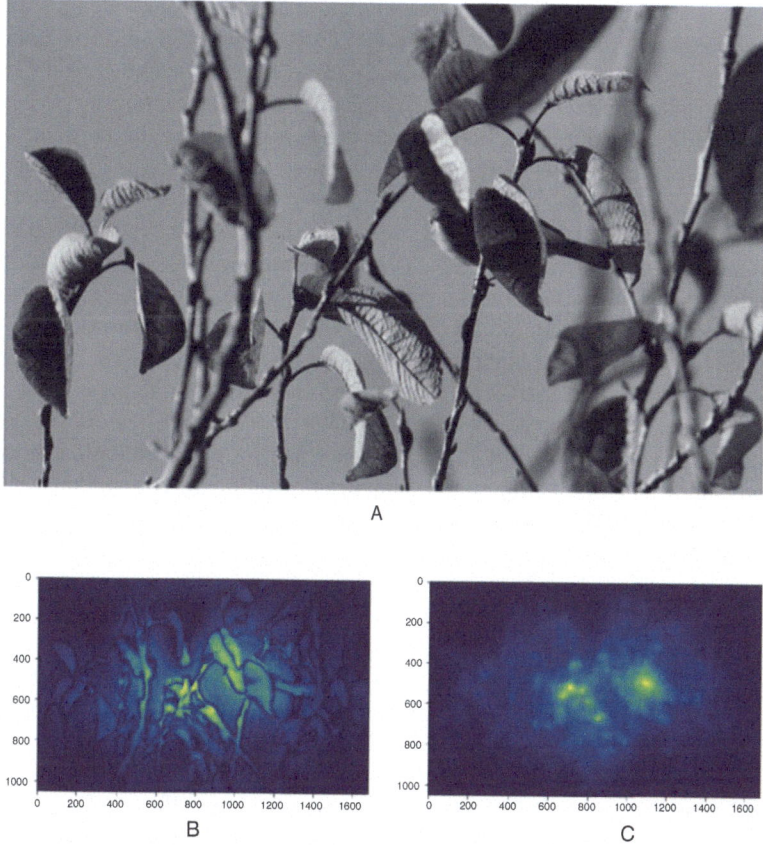

Fig. 5 (**a**) An image containing a bird, (**b**) the bottom-up saliency map, and (**c**) the top-down saliency map. Image Credit [6]

salient location in the visual field, ignoring irrelevant information and distractors, and allocating system processing resources for efficient and accurate processing.

There are two main ways to form the saliency map. "Top-down" attention is one way, and it is based on prior knowledge and our current goals. Bottom-up attention is the other way, based on how different something is compared with its neighbors. In the bottom-up attention, the saliency maps are calculated using multiple weighted visual features, such as color, texture, pixel intensity, edges, motion, etc. Figure 5 illustrates an example of a top-down and bottom-up saliency map.

Asgari et al. [4] have proposed a new digital architecture implementation of a saliency-based selective visual attention model. Here, saliency is defined as the region with the highest concentration of high-contrast moving objects. This is because the pixels in the dynamic vision sensor (DVS) primarily respond to local changes in the light intensity, effectively acting as local edge detectors for detecting moving objects. In this implementation, a window centered around the most salient pixel is defined as the Focus of Attention (FOA). Bottom-up selective attention models are suitable for hardware implementations because they are modular and easy to expand. However, in the implementation before the bottom-up processing stage, top-down biasing is applied to steer the FOA based on the target task [4]. There are two methods of doing top-down biasing. One method is known as top-down gating, which removes all areas outside of the predefined region of interest. Furthermore, top-down modulation is the second method that processes all pixels with a modulated state-changing gain. This acts as an amplitude modulation, where the amplitude of the modulation changes according to the predefined conditions.

After creating the saliency map, Winner-Take-All (WTA) network is used to select the location with the highest value in the saliency map, thereby determining the focus of attention. The WTA mechanism ensures that only one location is selected as the current focus of attention. After processing a region, the region is suppressed using an inhibition of return mechanism. This inhibition of return prevents immediate reselection of the same location, allowing the next most salient location to win the WTA competition and become the new focus of attention.

The authors in [4] have successfully implemented a prototype version of an event-driven saliency-based attention model on an FPGA. The use of FPGA technology allows for rapid prototyping and real-time processing of the data generated by the silicon retina device. As shown in Fig. 6, the proposed hardware architecture comprises seven main functional blocks: Flip-flop Meta-stability Synchronizer (FMS), Handshake Receiver (HSR), Data Processing Element in Exploring Mode [DPE (Exp)], Top-Down Gating (TDG), Top-Down Modulation (TDM), Saliency Block (SAL), and Data Processing Element in Fovea Mode [DPE (Fov)].

FMS is used to prevent the metastable states caused by different clock domains at the interface. Stabilized inputs are fed into HSR, which handles the input events while maintaining a handshake protocol between receiver and sender blocks. The DVS camera serially sends the 2-D cartesian coordinates of the events. Therefore, DPE constructs the 2D format of the event data and generates the "pixel_ID" as a result. The entire input resolution is fed into the DPE (Exp) block, where the "pixel_ID" is extracted from the input events. Once extracted, the "pixel_ID" is then forwarded to the SAL block. DPE (Fov) block is used to find the events inside

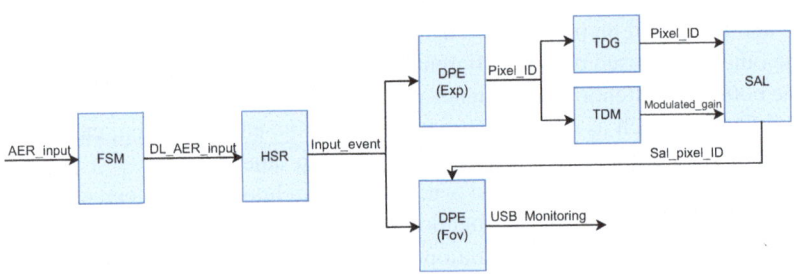

Fig. 6 Block diagram of the FPGA implementation of event-driven saliency-based attention model in [4]

the FOA by using the coordinate of the most salient pixel forwarded from the SAL block. The "pixel_ID," extracted using DPE (Exp), is used for the top-down biasing stage. This happens inside the TDG, which uses the top-down gating mechanism, and TDM, which uses the top-down modulation mechanism.

The most important block in the proposed architecture is the SAL block, where the most salient pixel is defined. Inside the SAL block, there are two dual-port BRAMs and two different functional units. The BRAMs store the pixel state and the timestamp of the most recent event recorded at each pixel location. When there is a new event from a particular pixel, the new state is computed by the functional block "Exp_Func," where the exponential decay operation is implemented using the last pixel event timestamp and the state. The new state and the timestamp are stored in BRAMs. The computed new state is checked with the state of the current salient pixel. If the new state value of a new pixel is found to be greater than the state of the current salient pixel, the location of the most salient pixel is updated with the coordinates of the new pixel. This updated salient pixel location is then forwarded to the DPE (Fov) block for further processing. In the final step, the Inhibition of Return (IOR) mechanism is employed to suppress the salient region that was previously selected.

The authors conducted the experiment using a Kintex-7 XC7KT160T FPGA. As discussed earlier, this model is specifically designed for the resource-constrained environment, and the experimental results demonstrate the feasibility of the model. Only 1% of the available slice flip-flops, DSPs, and LUTRAMs were utilized to implement the model while maintaining a latency of the system of just a few microseconds. The authors have chosen the DVS128 Gesture Dataset to validate the ability to find the trajectory of the most saliency locations of the proposed mode with real-world hand gestures under different illumination conditions. Additionally, an experiment with three active pixels with different firing rates was conducted to validate the effectiveness of the inhibition of return mechanism. Furthermore, another experiment explored the top-down biasing conditions' effect on TDM and TDG. Overall, the experimental results provide compelling evidence supporting the effectiveness and applicability of the proposed event-based saliency attention model with top-down modulation. Although the implementation of a saliency-based selective visual attention model is not an accelerator designed for an explainable AI model, the implementation of the saliency map can be used in an explainable AI accelerator.

4 GPU-Based Acceleration of Shapley Value Analysis

The discussion in this section is based on the work of Mitchell et al. [13]. Despite the importance of the interpretability of ML models, interpreting complex models is a very challenging problem. Tree ensemble attained using gradient boosting is one such complex model. A tree ensemble is used to get a better prediction than a single decision tree by combining several decision trees. Moreover, gradient boosting is a state-of-the-art ML technique that combines multiple weak models, usually decision trees, to create a strong model.

Local interpretability encompasses a set of methods to understand individual predictions made by ML models. Shapley additive explanation (SHAP) value is a result of one such method, which provides insights into the contribution of each feature toward the final decision. The concept underlying SHAP value is derived from the Shapley value, commonly used in game theory. Equation 6 shows the definition of SHAP values with respect to a decision tree model. In Eq. 6, $g\left(z'\right)$ represents a local explanation of the original trained decision tree model f for a given input instance x. The binary number z'_i indicates the presence of the ith feature among M number of features, and ϕ_i is defined as SHAP values that indicate the relative contribution of the given ith feature to the model decision. These SHAP values can be calculated using Eq. 7, where S is a feature subset. This equation calculates the average contribution of a given feature i by considering the weighted difference of outputs between a feature subset with the feature i and without feature i for all possible feature subsets. The weights of this equation depend on the cardinality of the subset S. The above consideration of all possible subsets leads to an exponential time calculation of SHAP values.

$$g\left(z'\right) = \phi_0 + \sum_{i=1}^{M} \phi_i z'_i \tag{6}$$

$$\phi_i = \sum_{S \subseteq M \setminus \{i\}} \frac{|S|!(|M| - |S| - 1)!}{|M|!} [f_{S \cup \{i\}}(x) - f_S(x)] \tag{7}$$

Also, there is no exact method to restrict model prediction $f(x)$ to feature subset S when f contains binary splits. The challenge arises because the decision of the split condition j becomes problematic when the feature j is not in the subset S. To address this issue, Lunderberg et al. [12] defined a conditional expectation for decision tree models, where the split condition j is treated as a Bernoulli random variable. When the split condition j is encountered, the outputs of both the left and right branches are considered. The estimated probability for the Bernoulli variable j for the left or right branch is determined during the model training as a proportion of the number of samples falling into left and right branches.

The introduction of conditional expectation for missing features led to the development of TreeShap, a polynomial-time algorithm proposed by Lundberg et al. [12] to solve Eq. 7. Equation 7 can be interpreted as the independent processing of every unique path from the root to the leaf in the decision tree. This algorithm uses a dynamic programming algorithm to process each unique path. It maintains information about the proportion of all possible subsets falling on each tree's leaves, all possible subset sizes, and estimated probabilities of missing features. The TreeShap algorithm exhibits a recursive tracking behavior, which allows it to efficiently compute the SHAP values.

Algorithm 1: TreeShap algorithm

Input   : x: input vector, $tree$: {v, a, b, t, r, d}
Output : ϕ: array of len(x) SHAP values
Procedure $TREESHAP\,(x, tree)$

1	Initiate array ϕ of size len(x) with all zeros
	Procedure $RECURSE\,(j, m, p_z, p_o, p_i)$
2	Use $EXTEND$ to extend the subset path
3	Check whether the current node is a leaf node
4	(*If YES:*)
5	(*LOOP to select every feature in the unique path*)
6	Use $UNWIND$ to undo the weight extension for the feature
7	Add the contribution from the subset to the feature's overall contribution
8	(*If NO:*)
9	Determine hot and cold children
10	Use $UNWIND$ to undo the previous feature extension if it is a duplicate
11	Use $RECURSE$ to send both zero and one weights to the hot child
12	Use $RECURSE$ to just send zero weight to the cold child
	Procedure $EXTEND\,(m, p_z, p_o, p_i)$
13	Initiate the subsets of size l
14	Grow subsets by one using p_o and keep subsets at the same size using p_z
15	**return** m
	Procedure $UNWIND\,(m, i)$
16	Do the inverse of the ith call to $EXTEND$
17	**return** m
18	Start $RECURSE$ at the first node with all zero and one extensions
19	**return** ϕ

The RECURSE method of Algorithm 1 maintains the recursive behavior of the algorithm. The method EXTEND is used to increase the feature subset sizes based on the provided p_o and p_z values, where p_o and p_z are the fractions of ones and zeroes that are going to extend the subsets, respectively. When invoking the EXTEND method, a new split is added to the current path of unique features that have been split so far (m). Also, this operation modifies the attribute w, which indicates the weighted number of already created feature subsets as a proportion of the total number of the possible subsets of a given cardinality. In addition to the attribute w, each element in m has three other attributes: d, z, and o, which represent the feature index, the fraction of (zeros) paths flow through when the feature is not in the set S, and the fraction of (ones) paths flow through when the feature is in the set S, respectively. The value p_i is another input for the EXTEND method, which represents the feature index of the last split.

The UNWIND method is used as an undo operation of the EXTEND method. Its primary purpose is twofold: to avoid adding duplicate features and calculate the correct weights for SHAP values for each feature in the path when it reaches a leaf node. The set of vectors v, a, b, t, r, and d contain the decision tree information where each element represents a node. The vectors v, a, b, t, and r represent node values, left node indexes, right node indexes, split condition, and the number of samples falling into a given subtree, respectively. Let f be an ensemble of T decision trees, L the number of leaves, and D the maximum tree depth, and the TreeSHAP algorithm has a time complexity of $O\left(TLD^2\right)$. Furthermore, this implementation of TreeSHAP using Algorithm 1 has a memory complexity of $O\left(D^2 + M\right)$ to get above exponential time complexity improvement.

The SHAP interaction value is a more comprehensive explanatory metric than the SHAP value, giving insights into second-order connections between features. SHAP interaction value is described by the following equations, which are based on the Shapley interaction index from game theory. By observing the similarity between Eqs. 10 and 7, interaction values can be calculated by performing the TreeSHAP algorithm twice, with the condition of feature j present and not present. This is achieved by not including the feature j into the path when it is encountered in a unique path. Then the resulting ϕ_i is modified according to the presence of the feature j. If the feature j is not present, ϕ_i is weighted with the conditional expectation. Conversely, if the feature j is present, the split condition on feature j is evaluated, and ϕ_i is discarded from the not taken path, for $i \neq j$,

$$\phi_{i,j} = \sum_{S \subseteq M \setminus \{i,j\}} \frac{|S|!(|M| - |S| - 2)!}{2(M-1)!} \nabla_{i,j}(S) \tag{8}$$

where

$$\nabla_{i,j}(S) = f_{S \cup \{i,j\}}(x) - f_{S \cup \{i\}}(x) - f_{S \cup \{j\}}(x) + f_S(x) \tag{9}$$

$$= f_{S \cup \{i,j\}}(x) - f_{S \cup \{j\}}(x) - [f_{S \cup \{i\}}(x) - f_S(x)] \tag{10}$$

for $i = j$,

$$\phi_{i,j} = \phi_i - \sum_{j \neq i} \phi_{i,j} \tag{11}$$

There are previous attempts to accelerate the TreeSHAP algorithm on a CPU with thread-level parallelism. These approaches commonly utilize one thread per row of the input matrix to achieve acceleration. However, when it comes to GPUs, direct implementation of TreeSHAP on GPU is not feasible. This is because the threads in a GPU are different from a CPU's. One reason for this difference is that CPUs are optimized for latency, and GPUs are optimized for throughput. Thus, several changes are required in TreeSHAP to implement it on a GPU to achieve desired acceleration. The new algorithm, called "GPUTreeSHAP," utilizes a group of threads to calculate the SHAP value of a particular unique path from root to the leaf in Algorithm 1. In GPUTreeSHAP, this group of threads calculates all the SHAP values for a unique path and evaluation instance pair in a single GPU kernel. The

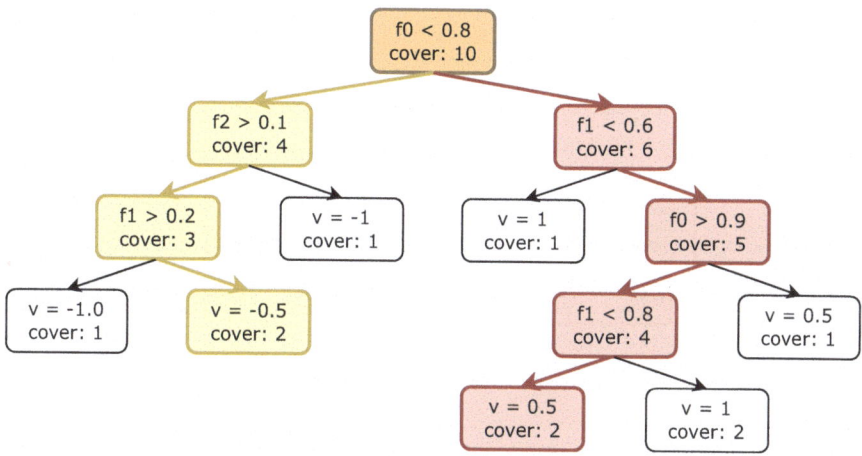

Fig. 7 An example decision tree model in [13]

abstract steps of GPUTreeSHAP are as follows:

Step 1: Arrange the tree ensemble to extract unique paths.
Step 2: Remove duplicate features along each path.
Step 3: Solve a bin packing problem to map path subproblems to GPU warps.
Step 4: Solve dynamic programming subproblems in a GPU kernel.

In the first step, the ensemble is preprocessed to generate a list of path elements. Each path element consists of the lower and upper bounds of a feature value when the feature is present, the fraction of paths that flow through when the feature is missing, and the leaf weight at the end of the path. Figure 7 depicts an example decision tree model, and Fig. 8 depicts two lists of path elements. Each path element contains information about its path index (path_idx; −1 is for root), the feature index of the split (feature_idx), the lower and upper bounds of the feature value that can flow through the selected path (lower_bound, upper_bound; inf is for infinity), the probability of following the path when the feature is not in the set (zero_fraction), and the leaf node weight (v). Although this new representation increases the space complexity, the experimental results indicate that the increment is negligible.

This new form of representation enables the ability to remove the redundancy of the same feature in a single unique path. This is an essential task because one of the reasons for the complexity of Algorithm 1 occurs due to the detection and handling of duplicate features in a unique path. In Algorithm 1, this is accomplished using the FINDFIRST and UNWIND functions. However, this duplicate finding and removing method is not feasible for GPU implementation due to branching and additional computation. With the help of new path element representation, a unique path can be depicted as a single hyperrectangle in the M-dimensional feature space. The boundaries of the hyperrectangle are established based on the split conditions encountered along the unique path and can be represented with lower and upper

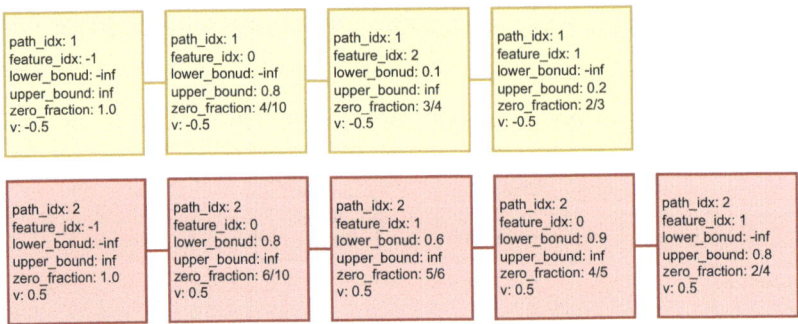

Fig. 8 An example list of path elements in [13]

bounds of feature value. As a result, multiple decision tree splits over a single feature can be reduced into a unified range. This reduction is possible because, as stated in [12], SHAP values are independent of the order of the features in a unique path. Therefore, a list of path elements of a particular unique path can be sorted by feature index, and duplicate occurrences of features can be combined into a single path element.

The next step is to allocate GPU warps for the above unique path subproblem. The objective of this task is to maximize the throughput of the GPU by effectively utilizing the SMs by keeping them busy with executing threads. To minimize the synchronization cost, the preference is to map a single unique path into threads in a single warp. If there are any unmapped threads from the 32 threads in the partially mapped warp, another suitable unique path should be mapped to maximize the utilization. As a result, a single warp can consist of multiple unique paths, but each unique path is only mapped into a single warp. To meet this constraint, the maximum depth of a decision tree should not exceed 32. Nevertheless, experimental results demonstrate that the maximum depth of real-world decision trees is almost always less than or equal to 16. However, for a given set of unique paths, finding the best allocation pattern that minimizes the number of warps is a bin packing problem, and solving a bin packing problem to find the optimal solution is NP-complete.

There are several standard heuristics that provide close to optimal solutions in a polynomial time. The authors have evaluated three such heuristics: Next-Fit (NF), First-Fit-Decrease (FFD), and Best-Fit-Decrease (BFD) [13]. The Next-Fit algorithm checks whether the current item fits in the current bin or not. If it fits, it is added there; otherwise, the current bin is closed, and a new bin is opened. Both First-Fit-Decreasing and Best-Fit-Decreasing algorithms first sort the item list in decreasing order. The FFD algorithm checks whether the item fits in any currently open bins or not. If it fits in multiple bins, the item is added to the first bin it fits. A new bin is opened if the item does not fit in any open bins. The BFD algorithm adds the item to the bin where it fits the tightest, and if it does not fit in any open bins, a new bin is opened. In the context of bin packing problems, the approximation ratio of any such algorithm is defined as the ratio between the number of bins used by the algorithm and the number of bins in the optimal solution.

As mentioned earlier, to gain the advantages of GPUs, the algorithm being implemented on a GPU should have parallelism, which is a challenging task for the above heuristics. Anderson et al. [2] have proposed an algorithm that has the same approximation ratio as FFD and BFD but can also parallelize. However, implementation of that algorithm in a GPU is also nontrivial. Fortunately, experimental results show that implementing NF, FFD, and BFD on a CPU can provide acceptable performance for the main problem. Although the CPU is used to solve the bin packing problem, the main processing of the SHAP value calculation is performed by GPU kernels. After the decision tree preprocessing and solving the bin packing problem, GPU threads are mapped with path elements. These threads compute all SHAP values for a given unique path and evaluation instance in a single GPU kernel. This single GPU kernel is the basis of GPUTreeSHAP implementation, which enables parallelization and allows multiple threads to work simultaneously on different parts of the Shapley value calculation concerning dataset rows, unique paths, and elements within each unique path.

Although the authors altered the unwinding function in the duplicate elimination step, there is another place that uses the unwinding function to calculate the relative contribution of each feature in the unique path. Additionally, the EXTEND function, which is the opposite of UNWIND, requires a new GPU-based implementation. Note that the EXTEND function is a single step in a dynamic programming problem. A single unique path is assigned to threads in a single warp. Therefore, when each thread handles a single path element, data dependencies arise between the threads, leading to the data dependency pattern in the execution of EXTEND function, as depicted in Fig. 9. In this figure, each thread depends only on two thread values: its own previous result and the previous result of its left thread. The warp shuffle

Fig. 9 Data dependency
pattern in EXTEND function

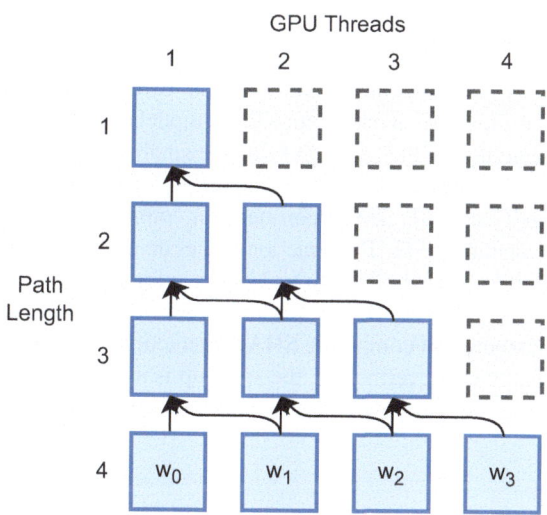

instruction is a perfect match for this implementation, as it can directly access other threads' registers in the same warp. A similar approach can be used to undo the effect of EXTEND function for a given feature to evaluate the relative contribution of that feature. The result of this new UNWIND function, which replaces line 6 in Algorithm 1, is used to calculate the final SHAP value.

Since the computation of SHAP interaction values uses the same TreeSHAP algorithm with slight modifications, a similar approach can be applied to the computation of SHAP interaction values using GPUTreeSHAP. The modification involves calculating SHAP values for each unique feature while considering it fixed as either present or absent in the model. However, implementing the above SHAP interaction value calculation with the discussed GPUTreeSHAP is difficult due to the conditioning on features. Nevertheless, by rearranging the path elements index, where the path element on which the condition is applied is swapped to the end of the path, we can simply avoid adding that feature to the list. Then, the SHAP interaction values can be calculated using a similar GPU kernel as in GPUTreeSHAP implementation and iterating over every unique feature with the condition of that feature being on or off.

For the experiments, the authors have used the XGBoost algorithm to train several tree ensembles on different datasets, aiming to cover a wide range of decision tree models. The experiments were conducted to evaluate the bin packing performance, SHAP value throughput, and SHAP interaction values. The experiment on bin packing performance involved a comparison of the NF, FFD, and BFD bin packing heuristics with a no-packing situation, considering the aspects such as execution time, utilization, and the used number of bins. According to the experimental results [13], FFD and BFD demonstrated better utilization than NF. However, in terms of execution time, NF outperformed both FFD and BFD. Based on the overall results, BFD was selected for implementation due to its ease of implementation, high utilization, acceptable execution time, and, most notably, its high approximation ratio. The experiment on GPUTreeSHAP as a backend to the XGBoost library compared its execution time with a CPU implementation of the TreeSHAP algorithm. The results show that the GPUTreeSHAP on a single V100 GPU outperformed TreeSHAP on 40 CPU cores, showcasing speedups of 13 to 19 times for medium and large models. Although TreeSHAP has less execution time than GPUTreeSHAP for the small models due to the high latency associated with launching the GPU kernels, GPUTreeSHAP has a higher throughput compared to TreeSHAP. The experiment on comparing GPU vs. 40-core CPU runtimes for calculating SHAP interaction values demonstrates that the GPU implementation can achieve speedups of up to 340 times for certain datasets due to the high throughput of the GPU and the algorithmic optimization. The experimental results highlight the challenge of computing SHAP interaction values for datasets with a large number of features. Nevertheless, the speedup is approximately comparable when the dataset contains a low number of features.

4.1 Summary

In this chapter, we surveyed hardware acceleration of explainable AI algorithms. Specifically, we described three hardware-based acceleration frameworks: an FPGA-based heatmap visualization accelerator, an FPGA-based model distillation accelerator, and a GPU-based Shapley value analysis accelerator. The experimental results demonstrated significant speedup compared to software implementation. Clearly, hardware acceleration can lead to utilization of explainable AI in real-time systems.

References

1. Zhixin Pan and Prabhat Mishra. Accelerating spectral normalization for enhancing robustness of deep neural networks. In *IEEE Computer Society Annual Symposium on VLSI, ISVLSI 2021, Tampa, FL, USA, July 7-9, 2021*, pages 260–265. IEEE, 2021.
2. Richard J Anderson, Ernst W Mayr, and Manfred K Warmuth. Parallel approximation algorithms for bin packing. *Information and Computation*, 82(3):262–277, 1989.
3. Daniel Arp et al. Effective and efficient malware detection at the end host. In *18th USENIX*, Montreal, Quebec, 2009.
4. Hajar Asgari, Nicoletta Risi, and Giacomo Indiveri. FPGA implementation of an event-driven saliency-based selective attention model. In *2022 IEEE Biomedical Circuits and Systems Conference (BioCAS)*, pages 307–311. IEEE, 2022.
5. Ashwin Bhat, Adou Sangbone Assoa, and Arijit Raychowdhury. Gradient backpropagation-based feature attribution to enable explainable-AI on the edge. In *2022 IFIP/IEEE 30th International Conference on Very Large Scale Integration (VLSI-SoC)*, pages 1–6. IEEE, 2022.
6. Ryan Burt, Nina N Thigpen, Andreas Keil, and Jose C Principe. Unsupervised foveal vision neural networks with top-down attention. *arXiv preprint arXiv:2010.09103*, 2020.
7. Javier Civit-Masot et al. TPU cloud-based generalized u-net for eye fundus image segmentation. *IEEE Access*, 7:142379–142387, 2019.
8. George E. Dahl, Jack W. Stokes, Li Deng, and Dong Yu. Large-scale malware classification using random projections and neural networks. In *ICASSP*, pages 3422–3426, 2013.
9. Kathrin Grosse et al. Adversarial perturbations against deep neural networks for malware classification. *CoRR*, 2016.
10. Tianjian Lu et al. Large-scale discrete Fourier transform on TPUs. *CoRR*, abs/2002.03260, 2020.
11. Tianjian Lu, Thibault Marin, Yue Zhuo, Yi-Fan Chen, and Chao Ma. Accelerating MRI reconstruction on TPUs. *CoRR*, abs/2006.14080, 2020.
12. Scott M Lundberg, Gabriel Erion, Hugh Chen, Alex DeGrave, Jordan M Prutkin, Bala Nair, Ronit Katz, Jonathan Himmelfarb, Nisha Bansal, and Su-In Lee. From local explanations to global understanding with explainable AI for trees. *Nature Machine Intelligence*, 2(1):56–67, 2020.
13. Rory Mitchell, Eibe Frank, and Geoffrey Holmes. GPUTreeShap: massively parallel exact calculation of SHAP scores for tree ensembles. *PeerJ Computer Science*, 8:e880, 2022.
14. Zhixin Pan and Prabhat Mishra. Automated detection of Spectre and Meltdown attacks using explainable machine learning. In *2021 IEEE International Symposium on Hardware Oriented Security and Trust (HOST)*, pages 24–34. IEEE, 2021.

15. Zhixin Pan and Prabhat Mishra. Hardware acceleration of explainable machine learning. In *2022 Design, Automation & Test in Europe Conference & Exhibition (DATE)*, pages 1127–1130. IEEE, 2022.
16. Zhixin Pan and Prabhat Mishra. A survey on hardware vulnerability analysis using machine learning. *IEEE Access*, 10:49508–49527, 2022.
17. Zhixin Pan and Prabhat Mishra. AI Trojan attack for evading machine learning-based detection of hardware Trojans. *IEEE Transactions on Computers*, 2023.
18. Zhixin Pan, Jennifer Sheldon, and Prabhat Mishra. Hardware-assisted malware detection and localization using explainable machine learning. *IEEE Transactions on Computers*, 71(12):3308–3321, 2022.
19. Joshua Saxe and Konstantin Berlin. Deep neural network based malware detection using two dimensional binary program features. In *10th MALCON*, pages 11–20, 2015.
20. Alan Tatourian. NVIDIA GPU architecture and CUDA programming environment. https:// tatourian.blog/2013/09/03/nvidia-gpu-architecture-cuda-programming-environment/, 2013.
21. Qinglong Wang et al. Adversary resistant deep neural networks with an application to malware detection. In *Proceedings of the 23rd ACM SIGKDD*, pages 1145–1153, 2017.
22. Hasini Witharana and Prabhat Mishra. Speculative load forwarding attack on modern processors. In *IEEE/ACM International Conference on Computer-Aided Design*, pages 1–9, 2022.
23. Zhixin Pan and Prabhat Mishra. Hardware Trojan detection using Shapley ensemble boosting. pages 1127–1130, 2021.

Explainable AI Acceleration Using Tensor Processing Units

1 Introduction

The advances in machine learning (ML) algorithms have enabled promising performance with outstanding flexibility [2, 3, 5, 13, 20, 22] and generalization [1, 11, 15, 16, 23] as demonstrated in the previous chapters. However, most of the existing ML methods are not able to interpret the outcome (e.g., explain its prediction) since it produces the outcome based on computations inside a "black-box." Explainable AI provides interpretation of input–output mapping as well as clues for importance ranking of input features [12, 24]. In other words, explainable AI acts like a supervisor to guide the learning process and provides additional information to users. Figure 1 shows that explainable AI has been used in a wide variety of application domains. For example, during the training of a facial recognition classifier, if we can obtain information about which region in the face distinguishes the target from the others, the corresponding weight can be adjusted to emphasize features collected from that region.

Due to the inherent inefficiency in explainable AI algorithms, they are not applicable in real-time systems. Specifically, it solves a complex optimization problem that consists of numerous iterations of time-consuming computations. As a result, such time-consuming interpretation is not suitable for time-sensitive applications with soft or hard deadlines. In soft real-time systems, such as multimedia and gaming devices, inefficient interpretation can lead to unacceptable Quality-of-Service (QoS). In hard real-time systems, such as safety-critical systems, missing task deadlines can lead to catastrophic consequences.

In this chapter, we discuss an efficient framework to achieve fast explainable AI utilizing Tensor Processing Units (TPUs). Specifically, we transform explainable AI procedure to computation of matrix operations and exploit the natural ability

Z. Pan, P. Mishra, *Explainable AI for Cybersecurity*,
https://doi.org/10.1007/978-3-031-46479-9_11

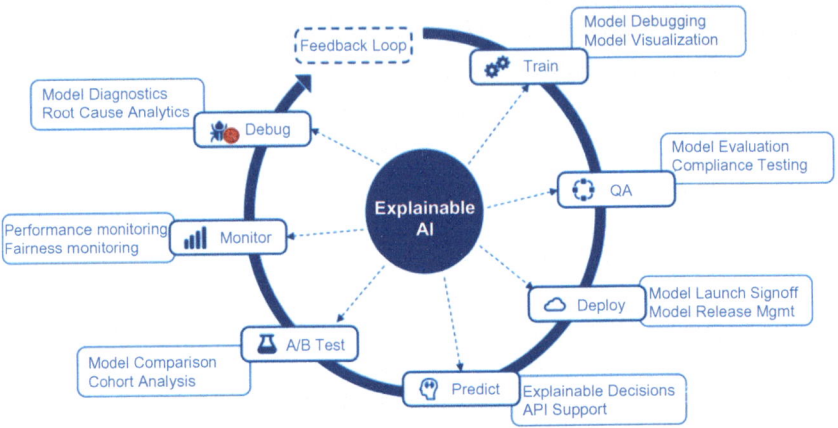

Fig. 1 Wide adoption of explainable machine learning. Image credit: [4]

of hardware accelerators for fast and efficient parallel computation of matrix operations. We first provide an overview of TPU architecture. Next, we describe TPU-based acceleration of explainable AI. Finally, we present some experimental results.

2   Tensor Processing Units

Tensor processing unit (TPU) is a domain-specific hardware for accelerating computation process of deep learning models. There are two fundamental reasons for the superior performance of TPUs: *quantization* and *systolic array* [21]. Quantization is the first step of optimization, which uses 8-bit integers to approximate 16-bit or 32-bit floating-point numbers. This can reduce the required memory capacity and computing resources. Systolic array is a major contributor to TPU's efficiency due to its natural compatibility with matrix manipulation coupled with the fact that computation in neural networks can be represented as matrix operations.

Figure 2 shows an overview of a TPU architecture. The core of the entire TPU is the matrix multiply unit, which is a 256×256 systolic array composed of multiple computation cells. Each cell receives a weight parameter along with an input signal at a time and performs accumulation of their products. Once all weights and input signals are propagated to neighboring cells, top to bottom and left to right, respectively, it immediately starts the next round of computations. As a result, the entire matrix multiplication can be completed by the collaboration of all computation cells. The systolic array of MXU contains $256 \times 256 = 65,536$ ALUs, which means that the TPU can process 65,536 8-bit integer multiplications

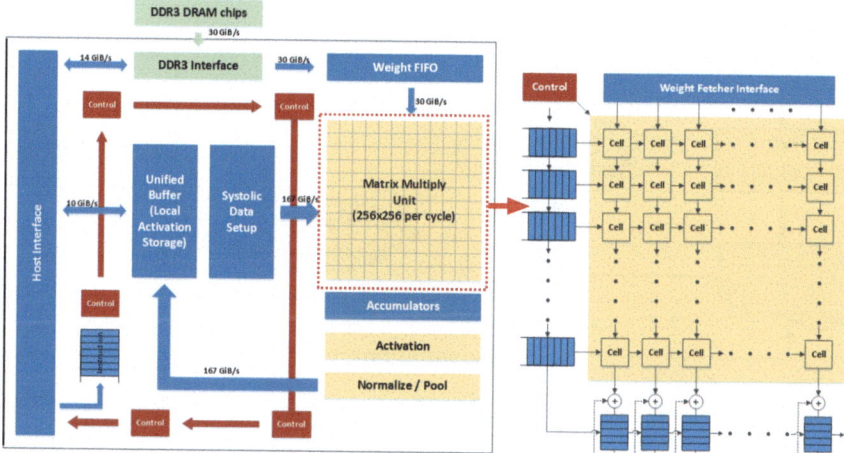

Fig. 2 An example TPU architecture where matrix multiply unit (MXU) is implemented by systolic array. Image Credit [10]

and additions per cycle. Due to the systolic architecture, input data can be reused for multiple times. Therefore, it can achieve higher throughput while consuming less memory bandwidth.

3 Hardware Acceleration of Explainable AI

Figure 3 shows an overview of our proposed framework for hardware acceleration of explainable machine learning. For a specific ML task, we apply traditional training scheme to construct a well-trained model and respective input–output dataset. Then we build a corresponding distilled model, which is able to provide reasonable explanation for target model's behavior. In this chapter, we consider three major tasks to achieve fast model distillation. First, we perform *task transformation* to map the explanation problem to Fourier transform computation by utilizing the inherent property of matrix convolution (Sect. 3.1). Next, we develop two synergistic activities to accelerate the computation procedure of Fourier transform. The first activity performs *data decomposition* (Sect. 3.2), where the complete computing task is split into multiple sub-tasks, and each sub-task can be executed by a GPU or TPU core without requiring any data exchange between cores (sub-tasks). The second activity fully exploits hardware accelerators' inherent ability in *parallel computation* to process multiple input–output pairs concurrently. Simultaneous execution of these two activities can provide significant improvement in acceleration efficiency, which is demonstrated in our experimental evaluation (Sect. 4).

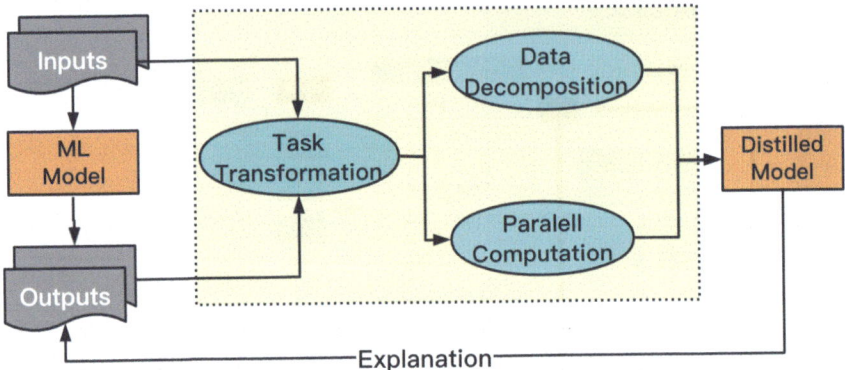

Fig. 3 Our framework consists of three major activities: task transformation, data decomposition, and parallel computation

3.1 Task Transformation

In this section, we demonstrate how to convert task into a matrix computation. *Model Specification:* In this chapter, a regression model is applied in order to satisfy the two requirements outlined above. Formally, given input data **X** and output **Y**, we find a matrix **K** using Eq. 1, where "*" denotes the matrix convolution.

$$\mathbf{X} * \mathbf{K} = \mathbf{Y} \tag{1}$$

Since convolution is a linear-shift-invariant operation, the above regression guarantees the distilled model to be sufficiently lightweight and transparent. Under this scenario, the model computation task boils down to solving for the parameters in matrix **K**.

Model Computation To solve for **K**, one key observation is that we can apply Fourier transformation on both sides of Eq. 1, and by discrete convolution theorem, it gives

$$\mathbf{X} * \mathbf{K} = \mathbf{Y}$$
$$\mathscr{F}(\mathbf{X} * \mathbf{K}) = \mathscr{F}(\mathbf{Y}) \tag{2}$$
$$\mathscr{F}(\mathbf{X}) \circ \mathscr{F}(\mathbf{K}) = \mathscr{F}(\mathbf{Y})$$

where ∘ is the Hadamard product. Therefore, the solution is given by this formula:

$$\mathbf{K} = \mathscr{F}^{-1}(\mathscr{F}(\mathbf{Y})/\mathscr{F}(\mathbf{X})) \tag{3}$$

Outcome Interpretation The primary goal of explainable ML is to measure how each input feature contributes to the output value. Once **K** is obtained, the contribution of each feature can be viewed in an indirect way—consider a scenario where we remove this component from the original input, and let it pass through the distilled model again to produce a "perturbed" result. Then by calculating the difference between the original and newly generated outputs, the impact of the key feature on the output can be quantified. The intuition behind the assumption is that hiding important features are more likely to cause considerable changes to the model output. Formally, assume that the input is $\mathbf{X} = [\mathbf{x}_1, \mathbf{x}_2, \ldots, \mathbf{x}_{i-1}, \mathbf{x}_i, \mathbf{x}_{i+1} \ldots, \mathbf{x}_d]$. We define the contribution factor of $\mathbf{x}_i$ as

$$con(\mathbf{x}_i) \triangleq \mathbf{Y} - \mathbf{X}' * \mathbf{K} \tag{4}$$

where $\mathbf{X}' = [\mathbf{x}_1, \mathbf{x}_2, \ldots, \mathbf{x}_{i-1}, \mathbf{0}, \mathbf{x}_{i+1} \ldots, \mathbf{x}_d]$, which is nothing but removing the target component from the original input.

As we can see, the original model distillation task has been converted into a matrix computation problem, which consists of matrix convolution, point-wise division, and Fourier transform. The first two types of operations can be inherently accelerated by GPU or TPU's built-in structure [9]. The next section describes the details for accelerating Fourier transform.

3.2 Data Decomposition in Fourier Transform

In this section, we demonstrate how to apply data decomposition to disentangle Fourier transform computation and further utilize computation resources to significantly accelerate the computing process. The general form of a 2-D Discrete Fourier Transform (DFT) applied on an $M \times N$ signal is defined as

$$X[k, l] = \frac{1}{\sqrt{MN}} \sum_{n=0}^{N-1} \left[\sum_{m=0}^{M-1} x[m, n] e^{-j2\pi \frac{mk}{M}} \right] e^{-j2\pi \frac{nl}{N}} \tag{5}$$

where $k = 0, \ldots, M - 1, l = 0, \ldots, N - 1$.

If we define intermediate signal X' such that

$$X'[k, n] \triangleq \frac{1}{\sqrt{M}} \sum_{m=0}^{M-1} x[m, n] e^{-j2\pi \frac{mk}{M}} \tag{6}$$

and plug it into Eq. 5, we have

$$X[k, l] = \frac{1}{\sqrt{N}} \sum_{n=0}^{N-1} X'[k, n] e^{-j2\pi \frac{nl}{N}} \tag{7}$$

Notice the similarity to the definition of 1-D Fourier transform applied on a M-length vector:

$$X[k] = \frac{1}{\sqrt{M}} \sum_{m=0}^{M-1} x[m] e^{-j2\pi \frac{mk}{M}} \tag{8}$$

If we treat n as a fixed parameter, then application is equivalent to performing a 1-D Fourier transform on the n-th column of the original input $M \times N$ matrix. Note that for 1-D Fourier transform, it can always be written as a product of input vector and Fourier transform matrix.

This transformed expression indicates that a 2-D Fourier transform can be achieved in a two-stage manner. First, transform all the rows of matrix to obtain intermediate result. Second, transform all the columns of the resulting matrix. An important observation is that the required computation for each row/column is completely independent. This implies that in real implementation, we can always split the computation process into sub-threads.

Our analysis reveals that merging the results matches with the desired 2-D Fourier transform result. In addition to exploiting hardware to accelerate Fourier transform, we make use of parallel computing to further improve the time efficiency. Notice in the training phase, multiple inputs will be fed into the model to generate corresponding outputs. The above data decomposition technique is applied on each individual input such that the computation cost is distributed among several cores. Extending from single to multiple input is simple and only requires one-step further utilization of parallel computation.

An illustrative example is shown in Fig. 4 where the goal is to perform 1-D Fourier transform on each column of three input matrices. First, each input matrix is segmented into pieces, and each core obtains a slice of them. Next, each piece is assigned to an individual core to perform the Fourier transform. During computation, an internal table is utilized to keep track of the distribution to guide the process of reassembling. In terms of matrix multiplication, the framework is exactly the same except the fact that *block matrix multiplication* is applied. Original matrices are partitioned into small blocks. By performing multiplication between blocks and merging afterward, we achieve same level of parallel computing efficiency. Due to the data decomposition step applied in Sect. 3.2, the whole computing procedure contains Fourier transform, matrix multiplication, and point-wise division only, which indicates parallelism is maintained across the entire computation process.

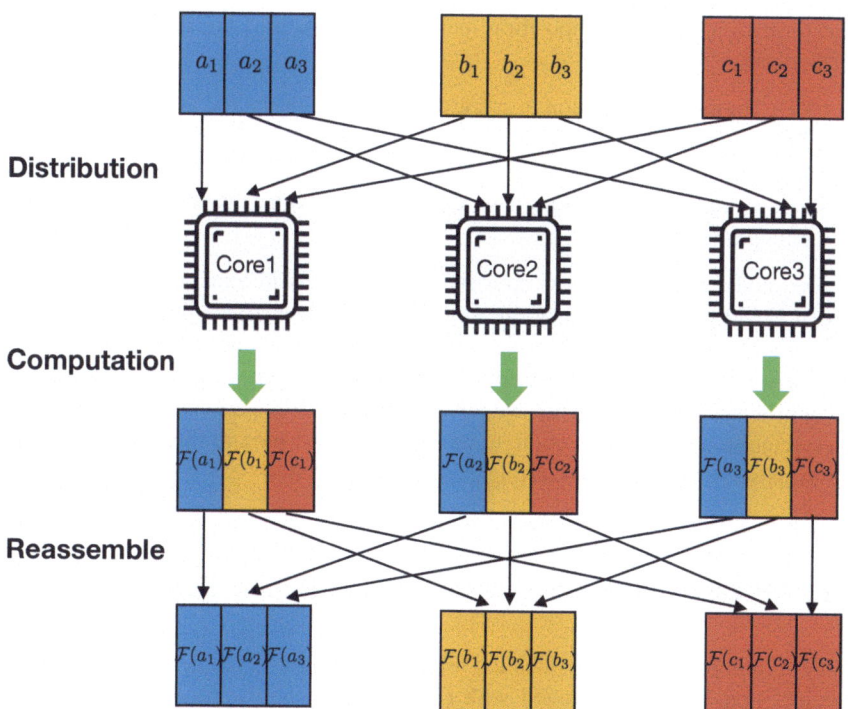

Fig. 4 An example of parallel computation in our framework. Each input is separated into pieces for multiple cores to run in parallel. The outputs are reassembled to obtain the results

4 Experiments

4.1 Experimental Setup

Experiments were conducted on a host machine with Intel i7 3.70 GHz CPU, equipped with an external NVIDIA GeForce RTX 2080 Ti GPU, which is considered as the state-of-the-art GPU accelerator for ML algorithms. We also utilize Google's Colab platform to access Google Cloud TPU service. In our evaluation, we used TPUv2 with 64 GB High Bandwidth Memory (HBM) and 128 TPU cores. We developed code using Python [6] for model training and PyTorch 1.6.0 [18] as the machine learning library. We have used the following two popular benchmarks in our study:

1. VGG19 [19] classifier for *CIFAR-100* classification
2. ResNet50 [7] network for malware detection [14, 17]

We have compared the following three configurations to highlight the importance of our proposed hardware acceleration approach. To address the compatibility of proposed optimization approach, all the proposed optimization methods (task transformation, data decomposition, parallel computation) are deployed on all 3 accelerators:

1. **CPU**: Traditional execution in software, which is considered as baseline method
2. **GPU**: NVIDIA GeForce 2080 Ti GPU, which is considered as state-of-the-art ML acceleration component
3. **TPU**: Google's cloud TPU, a specific ASIC designed to accelerate machine learning procedure

The model training process consists of 500 epochs in total, with a mini-batch size of 128. As for result evaluation, we first evaluated classification performance by reporting ML models' classification accuracy and execution time. Next, we compare the energy efficiency by measuring the performance per watt on each hardware under different workloads. Then we report the average time for completing outcome interpretation step for each configuration. Finally, we present the effectiveness of our proposed method in interpreting classification results.

4.2 Comparison of Accuracy and Classification Time

Table 1 compares the classification time and accuracy. Each row represents a specific model structure trained with corresponding hardware configuration. For both training time and testing time, each entry represents a time cost of 10 epochs on average. As we can see, with a sufficient number of training epochs, all methods obtain reasonable classification accuracy. However, when it comes to time efficiency, the CPU-based baseline implementation lags far behind the other two, which achieved the slowest speed. On VGG19, GPU provides the best acceleration performance, which provides 65x speedup compared to the baseline implementation. This clearly indicates the great compatibility between hardware accelerator and our proposed framework. In case of ResNet50, an even higher speedup was obtained by TPU, showing its acceleration potential in large-scale neural networks by providing around four times speedup than GPU. The drastic improvement (44.5x) compared to the baseline method also leads to significant energy savings, as described in the next section.

Table 1 Comparison of accuracy and classification time for various benchmarks

	CPU-based Acceleration			GPU-based Acceleration			TPU-based Acceleration		
Benchmark	Accuracy (%)	Training time(s)	Testing time(s)	Accuracy (%)	Training time(s)	Testing time(s)	Accuracy (%)	Training time(s)	Testing time(s)
VGG19	94.06	24.2	10.9	92.08	0.25	0.08	96.37	0.4	0.14
ResNet50	78.99	176.2	129.8	86.87	19.1	9.4	87.47	4.3	2.6
Average	86.52	100.2	70.35	89.47	9.67	4.84	91.92	2.35	1.37

4.3 Comparison of Energy Efficiency

Power consumption is another important aspect of performance evaluation, as power closely affects the thermal, provision, and stability of the device. Consequently, designers should consider power consumption for methods deployed on hardware components to ensure their power constraints are 5satisfied. Figure 5 shows the geometric and weighted mean performance/Watt for the RTX 2080 Ti GPU and Google's TPU relative to the CPU. Similar to [8], we calculate performance/Watt in two different ways. The first one (referred as "total") computes the total power consumption that consists of the power consumed by the host CPU as well as the actual execution performance/Watt for the GPU or TPU. The second one (referred as "incremental") does not consider the host CPU power, and therefore, it reflects the actual power consumption of the GPU or TPU during acceleration. As we can see from Fig. 5, for total performance/watt, the GPU implementation is 1.9X and 2.4X better than baseline CPU for geometric mean (GM) and weighted mean (WM), respectively. TPU outperforms both CPU (16x on GM and 33X on WM) and GPU (8.4X on GM and 13.8x on WM) in terms of total performance/watt. For incremental-performance/watt, when host CPU's power is omitted, the TPU shows its dominance in energy efficiency over both CPU (39x on GM and 69X on WM) and GPU (18.6x on GM and 31X on WM).

In terms of energy efficiency, TPU outperforms GPU, which outperforms CPU. Our proposed method fully utilizes data decomposition to create a high-level parallel computing environment where both GPU and TPU benefit from it to balance the

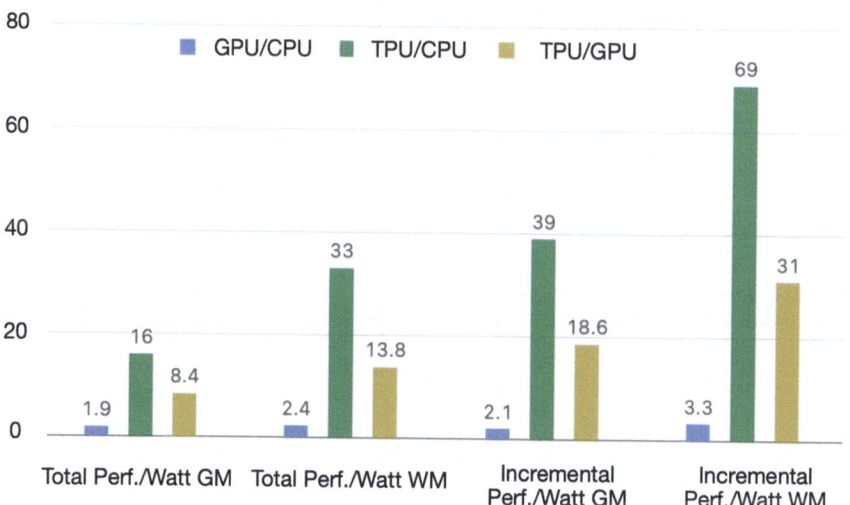

Fig. 5 Relative performance/Watt of GPU (blue bar) and TPU (green bar) over CPU, and TPU versus GPU (yellow bar). The total perf./watt includes host CPU power, while incremental ignores it. **GM** and **WM** are the geometric mean and weighted means, respectively

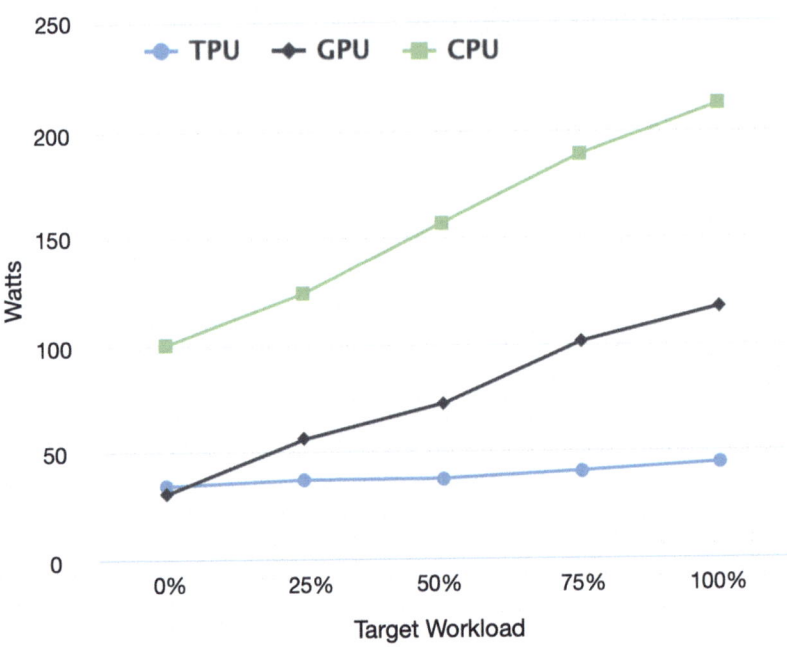

Fig. 6 The power consumption of different hardware with various workload situations

workloads on every single core. Although both GPU and TPU have the advantage of utilizing parallel computing to fulfill the proposed framework, TPU provides better performance/watt primarily due to the "quantification" property of TPU. Quantification is a powerful mechanism to reduce the cost of neural network prediction as well as the reduction in memory. The use of integers instead of floating-point calculations greatly reduces the hardware size and power consumption of the TPU. Specifically, TPU can perform 65,536 8-bit integer multiplications in a cycle, while mainstream GPUs used in cloud environments can perform few thousands of 32-bit floating-point multiplications. As long as 8 bits can be used to meet the accuracy requirements, it can bring significant performance improvement. While both TPU- and GPU-based acceleration can achieve fast explainable machine learning, the TPU-based implementation is the best in terms of energy efficiency, as shown in Fig. 6.

5   Summary

Long running time of explainable AI algorithms severely restricts their applicability in many domains, such as real-time systems. In this chapter, we address this fundamental bottleneck using hardware-based acceleration to provide explainability (transparency) of machine learning models in a reasonable time. Specifically, we described an efficient mechanism to transform the model distillation problem to linear algebra computation problem. The transformed model is able to fully exploit the inherent ability of hardware accelerators in computing ultra-fast matrix operations. Moreover, it enables parallel computing by performing data decomposition to break a large matrix into multiple small matrices. Experimental evaluation on a diverse set of benchmarks demonstrated that such an approach is scalable and able to meet real-time constraints. Our studies reveal that our proposed framework can effectively utilize the inherent advantages of both TPU- and GPU-based architectures. Specifically, TPU-based acceleration provides drastic improvement in interpretation time (39x over CPU and 4x over GPU) as well as energy efficiency (69x over CPU and 31x over GPU) for both image classification and malware detection benchmarks.

References

1. Alif Ahmed, Yuanwen Huang, and Prabhat Mishra. Cache reconfiguration using machine learning for vulnerability-aware energy optimization. *ACM Transactions on Embedded Computing Systems (TECS)*, 18(2):1–24, 2019.
2. Daniel Arp et al. Effective and efficient malware detection at the end host. In *18th USENIX*, Montreal, Quebec, 2009.
3. George E. Dahl, Jack W. Stokes, Li Deng, and Dong Yu. Large-scale malware classification using random projections and neural networks. In *ICASSP*, pages 3422–3426, 2013.
4. Krishna Gade, Sahin Cem Geyik, Krishnaram Kenthapadi, Varun Mithal, and Ankur Taly. Explainable AI in industry. In *Proceedings of the 25th ACM SIGKDD International Conference on Knowledge Discovery & Data Mining*, pages 3203–3204, 2019.
5. Kathrin Grosse et al. Adversarial perturbations against deep neural networks for malware classification. *CoRR*, 2016.
6. Kaiming He, Xiangyu Zhang, Shaoqing Ren, and Jian Sun. Deep residual learning for image recognition. *CoRR*, abs/1512.03385, 2015.
7. Kaiming He, Xiangyu Zhang, Shaoqing Ren, and Jian Sun. Deep residual learning for image recognition. *CoRR*, abs/1512.03385, 2015.
8. Norman P Jouppi, Cliff Young, Nishant Patil, David Patterson, Gaurav Agrawal, Raminder Bajwa, Sarah Bates, Suresh Bhatia, Nan Boden, and Al Borchers. In-datacenter performance analysis of a tensor processing unit. In *ISCA*, pages 1–12, 2017.
9. Cliff Young Kaz Sato and David Patterson. An in-depth look at Google's first tensor processing unit (TPU). 2017.
10. Tianjian Lu, Thibault Marin, Yue Zhuo, Yi-Fan Chen, and Chao Ma. Large-scale discrete Fourier transform on TPUs. *CoRR*, abs/2002.03260, 2020.
11. Zhixin Pan and Prabhat Mishra. Automated detection of Spectre and Meltdown attacks using explainable machine learning. In *2021 IEEE International Symposium on Hardware Oriented Security and Trust (HOST)*, pages 24–34. IEEE, 2021.
12. Zhixin Pan and Prabhat Mishra. Automated test generation for hardware Trojan detection using reinforcement learning. page 408–413, 2021.
13. Zhixin Pan and Prabhat Mishra. A survey on hardware vulnerability analysis using machine learning. *IEEE Access*, 10:49508–49527, 2022.
14. Zhixin Pan, Jennifer Sheldon, and Prabhat Mishra. Hardware-assisted malware detection using explainable machine learning. In *ICCD*, pages 663–666, 2020.

15. Zhixin Pan, Jennifer Sheldon, and Prabhat Mishra. Test generation using reinforcement learning for delay-based side-channel analysis. In *IEEE/ACM International Conference on Computer Aided Design (ICCAD)*, pages 1–7, 2020.
16. Zhixin Pan, Jennifer Sheldon, and Prabhat Mishra. Hardware-assisted malware detection and localization using explainable machine learning. *IEEE Transactions on Computers*, 71(12):3308–3321, 2022.
17. Zhixin Pan, Jennifer Sheldon, Chamika Sudusinghe, Subodha Charles, and Prabhat Mishra. Hardware-assisted malware detection using machine learning. In *Design Automation and Test in Europe (DATE)*, 2021.
18. A Paszke, S Gross, F Massa, A Lerer, Jea PyTorch Bradbury, G Chanan, T Killeen, Z Lin, N Gimelshein, and L Antiga. PyTorch: An imperative style, high-performance deep learning library. In *NIPS*, pages 8024–8035. 2019.
19. Hussam Qassim, David Feinzimer, and Abhishek Verma. Residual squeeze VGG19. *CoRR*, abs/1705.03004, 2017.
20. Joshua Saxe and Konstantin Berlin. Deep neural network based malware detection using two dimensional binary program features. In *10th MALCON*, pages 11–20, 2015.
21. Ehud Shapiro. Systolic programming: A paradigm of parallel processing. In *FGCS*, pages 458–470, 1984.
22. Qinglong Wang et al. Adversary resistant deep neural networks with an application to malware detection. In *Proceedings of the 23rd ACM SIGKDD*, pages 1145–1153, 2017.
23. Hasini Witharana and Prabhat Mishra. Speculative load forwarding attack on modern processors. In *IEEE/ACM International Conference on Computer-Aided Design*, pages 1–9, 2022.
24. Zhixin Pan and Prabhat Mishra. Hardware Trojan detection using Shapley ensemble boosting. pages 1127–1130, 2021.

Part VI
Conclusion

The Future of AI-Enabled Cybersecurity

1 Introduction

Our daily activities involve frequent interactions with the electronic systems, such as desktops, laptops, and smartphones. Security and privacy of such interactions rely on the trustworthiness of the underlying computing systems, which consists of hardware and software. In order to design trustworthy computing systems, we need to first identify the sources of hardware and software vulnerabilities. Next, we need to implement effective detection and mitigation methods to defend against cybersecurity attacks. For example, supply chain vulnerabilities can lead to malicious implants (e.g., hardware Trojans), backdoor for information leakage, or other integrity issues [1, 3]. Similarly, malicious software (e.g., malware and ransomware) is a serious threat to modern computing systems [12]. Clearly, there is an urgent need to develop efficient vulnerability detection and mitigation techniques to enable trustworthy computing.

2 Summary

This book provided a comprehensive overview of state-of-the-art vulnerability detection and mitigation methods using explainable AI. The previous chapters covered a wide variety of techniques to defend against security attacks. The topics covered in this book can be broadly divided into the following five categories.

© The Author(s), under exclusive license to Springer Nature Switzerland AG 2023
Z. Pan, P. Mishra, *Explainable AI for Cybersecurity*,
https://doi.org/10.1007/978-3-031-46479-9_12

2.1 Introduction to Cybersecurity and Explainable AI

The first two chapters introduced cybersecurity vulnerabilities and explainable AI algorithms.

- Chapter "Cybersecurity Landscape for Computer Systems" provided an overview of hardware and software vulnerabilities, such as hardware Trojans, information leakage, reverse engineering, malware, ransomware, etc. It also described recent approaches for defending against cybersecurity attacks using machine learning, simulation-based validation, and side-channel analysis [10].
- Chapter "Explainable Artificial Intelligence" introduced various machine learning algorithms, such as decision trees, random forest, reinforcement learning, and neural networks. It also described popular explainable AI methods, including model interpretability, knowledge extraction, saliency maps, integrated gradients, Shapley value analysis, and layer-wise relevance propagation.

2.1.1 Detection of Software Vulnerabilities

The next two chapters focused on the detection of malicious software attacks using explainable AI.

- Chapter "Malware Detection Using Explainable AI" presented a malware detection and localization framework using explainable AI. Specifically, it demonstrated that the explainable outcome through effective utilization of hardware performance counters and embedded trace buffer can lead to accurate localization of malicious behaviors [12].
- Chapter "Spectre and Meltdown Detection Using Explainable AI" described an efficient framework for the detection and localization of Spectre and Meltdown attacks using explainable AI. Specifically, it explored model distillation and the Shapley value analysis to improve the model's explainability. It also utilized long short-term memory and ensemble boosting to accelerate the training speed [5].

2.2 Detection of Hardware Vulnerabilities

The next three chapters deal with the detection of hardware Trojans using a wide variety of machine learning algorithms.

- Chapter "Hardware Trojan Detection Using Reinforcement Learning" described hardware Trojan detection using reinforcement learning. Specifically, it utilizes logic testing approach for Trojan detection using an effective combination of testability analysis and reinforcement learning [6].
- Chapter "Hardware Trojan Detection Using Side-Channel Analysis" provided an overview of hardware Trojan detection techniques using side-channel analysis.

Specifically, it utilized critical path analysis to generate test vectors that can maximize the side-channel sensitivity for delay-based detection of hardware Trojans [2, 11].

- Chapter "Hardware Trojan Detection Using Shapley Ensemble Boosting" utilized Shapley ensemble boosting for the detection of hardware Trojans. Specifically, it performed the Shapley value analysis to compute the importance ranking of input features to provide a guideline for feature selection and explainable interpretation. It also utilized boosting (ensemble learning) to generate a sequence of lightweight models to significantly reduce the training time [7].

2.3   Mitigation of AI Vulnerabilities

The next two chapters looked at mitigation of AI vulnerabilities in deep neural networks (DNNs).

- Chapter "Mitigation of Adversarial Machine Learning" explored the effectiveness of spectral normalization to design robust ML models against adversarial attacks. Specifically, it presented an acceleration technique for spectral normalization based on the Fourier transform and layer separation. The proposed method provides DNNs with promising security protection with minimal impact on time overhead [4].
- Chapter "AI Trojan Attacks and Countermeasures" presented various AI Trojan attacks and effective countermeasures. Specifically, it described a robust backdoor attack on ML-based Trojan detection algorithms. The proposed framework is able to design an AI Trojan and implant it inside the ML model that can be triggered by specific inputs [8].

2.4   Acceleration of Explainable AI

The next two chapters presented various methods for hardware-based acceleration of explainable AI using Field Programmable Gate Array (FPGA), Graphics Processing Units (GPUs), and Tensor Processing Unit (TPU).

- Chapter "Hardware Acceleration of Explainable AI" described hardware acceleration of explainable AI using FPGA and GPU. Hardware acceleration enables the fast and efficient explainable AI models that provide explainability and applicability across diverse domains, including real-time and safety-critical systems.
- Chapter "Explainable AI Acceleration Using Tensor Processing Units" presented TPU-based acceleration of explainable AI. Specifically, it explored the synergy between matrix convolution and Fourier transform, and therefore, it takes full advantage of TPU's inherent ability in accelerating matrix computations [9].

3 Future Directions

In spite of extensive research efforts in developing ML-based vulnerability detection techniques over the years, there are still many challenges that remain to design secure and trustworthy systems. In this section, we discuss few promising avenues to extend the proposed approaches to deal with future vulnerabilities.

3.1 Automatic Implementation of Secure Systems

There are many initiatives for an automatic implementation of secure components, such as secure silicon, secure compilation, etc. For example, the automatic implementation of secure silicon uses a combination of design time vulnerability analysis and runtime security monitoring techniques. Similarly, there are efforts to produce hardware and software components that can provide specific security guarantees. The future systems need to develop a framework to provide system-level security guarantees using component-level guarantees. There will also be a significant emphasis on detecting unintentional vulnerabilities. For example, existing Electronic Design Automation (EDA) tools and software compilers can introduce several types of security vulnerabilities. There should be significant investment in designing both EDA tools and compilers that are guaranteed to produce correct designs as per the specification—nothing more, nothing less. Clearly, someone has to deal with the hard task of developing a complete and accurate specification in the first place, and keep updating the specification when there are any subsequent changes.

3.2 Detection of Malicious Implants

The side-channel analysis is promising for detecting malicious implants using various side-channel signatures, including dynamic current, path delay, electromagnetic (EM) emanation, etc. This book explored test generation techniques to effectively maximize the sensitivity of path delay-based analysis. The concept of improving side-channel sensitivity can be explored for other possible side-channel signatures. For example, EM is closely related to the switching activity in the circuit. Therefore, the state transitions introduced by a Trojan can produce more electromagnetic emanations. This natural relationship between EM and switching activity makes it a promising signature for rare signal-based hardware Trojan detection.

3.3 Detection of Ransomware Attacks

This book described an efficient malware detection framework using explainable machine learning, which significantly improves the detection accuracy and transparency. However, this framework cannot be directly applied for ransomware detection. Although there are many ransomware detection methods that employ different techniques such as signature-based detection, honeypot, and hashing, these methods suffer from ransomware's two-stage strategy: (i) collects the user's data and registers for encryption algorithms and (ii) performs the actual attack (encrypting the user's files). If the detection takes effect in the first stage, the detection technique is less effective since it ignores the information provided in the second stage, leading to a high false positive rate. Once ransomware enters the second stage, the encryption process has already started. As a result, the effective detection of ransomware attack needs to balance between not too early (false positive) and not too late (a lot of files are encrypted already).

3.4 Automatic Data Augmentation

This book utilized explainable AI to defend against evasive Spectre and Meltdown attacks. The key idea is to apply adversarial training using synthetic samples. However, this data augmentation process is done manually, which introduces extra time overhead and must be guided by expert knowledge. To address this limitation, a natural idea is to enable automatic data augmentation utilizing various avenues, such as generative adversarial networks and diffusion models.

This book considered several types of hardware and software vulnerabilities. In the future, we have to deal with vulnerabilities that involve complex interactions of hardware, software, and firmware. We also have to deal with both known–unknown (e.g., known vulnerabilities with minor or major variations) as well as unknown–unknown (not even remotely related to any of known vulnerabilities) security attacks. There will be a need for developing hybrid approaches combining the inherent advantages of different vulnerability detection methods to detect a wide variety of security vulnerabilities in emerging computing systems.

References

1. Farimah Farahmandi, Yuanwen Huang, and Prabhat Mishra. *System-on-Chip Security: Validation and Verification.* Springer, 2020.
2. Yangdi Lyu and Prabhat Mishra. MaxSense: Side-channel sensitivity maximization for trojan detection using statistical test patterns. *TODAES*, 2020.
3. Prabhat Mishra, Swarup Bhunia, and Mark Tehranipoor. *Hardware IP security and trust.* Springer, 2017.

4. Zhixin Pan and Prabhat Mishra. Accelerating spectral normalization for enhancing robustness of deep neural networks. In *IEEE Computer Society Annual Symposium on VLSI (ISVLSI)*, pages 20–38, 2021.
5. Zhixin Pan and Prabhat Mishra. Automated detection of Spectre and meltdown attacks using explainable machine learning. In *2021 IEEE International Symposium on Hardware Oriented Security and Trust (HOST)*, pages 24–34. IEEE, 2021.
6. Zhixin Pan and Prabhat Mishra. Automated test generation for hardware trojan detection using reinforcement learning. In *Asia and South Pacific Design Automation Conference (ASPDAC)*, pages 408–413, 2021.
7. Zhixin Pan and Prabhat Mishra. Hardware trojan detection using Shapley ensemble boosting. pages 1127–1130, 2021.
8. Zhixin Pan and Prabhat Mishra. Design of ai trojans for evading machine learning-based detection of hardware trojans. In *Design Automation and Test in Europe (DATE)*, 2022.
9. Zhixin Pan and Prabhat Mishra. Hardware acceleration of explainable machine learning. In *Design Automation and Test in Europe (DATE)*, 2022.
10. Zhixin Pan and Prabhat Mishra. A survey on hardware vulnerability analysis using machine learning. *IEEE Access*, 10:49508–49527, 2022.
11. Zhixin Pan, Jennifer Sheldon, and Prabhat Mishra. Test generation using reinforcement learning for delay-based side-channel analysis. In *IEEE/ACM International Conference On Computer Aided Design (ICCAD)*, pages 1–7, 2020.
12. Zhixin Pan, Jennifer Sheldon, and Prabhat Mishra. Hardware-assisted malware detection and localization using explainable machine learning. *IEEE Transactions on Computers*, 71(12):3308–3321, 2022.

Index

© The Editor(s) (if applicable) and The Author(s), under exclusive license to Springer Nature Switzerland AG 2023
Z. Pan, P. Mishra, *Explainable AI for Cybersecurity*,
https://doi.org/10.1007/978-3-031-46479-9

GPSR Compliance

The European Union's (EU) General Product Safety Regulation (GPSR) is a set of rules that requires consumer products to be safe and our obligations to ensure this.

If you have any concerns about our products, you can contact us on ProductSafety@springernature.com

In case Publisher is established outside the EU, the EU authorized representative is:

Springer Nature Customer Service Center GmbH
Europaplatz 3
69115 Heidelberg, Germany

The manufacturer's authorised representative in the EU is Springer
Nature Customer Service Centre GmbH, Europaplatz 3, 69115 Heidelberg,
Germany. If you have any concerns regarding our products, please
contact ProductSafety@springernature.com

Printed and bound by CPI Group (UK) Ltd, Croydon, CR0 4YY

29/04/2026

02099548-0001